Igor Godinho Portis
Alex Lucas Hanusch
Cláudio Carlos Silva

Bioassays on the toxicogenetics of the fungicide Tebuconazole

Igor Godinho Portis
Alex Lucas Hanusch
Cláudio Carlos Silva

Bioassays on the toxicogenetics of the fungicide Tebuconazole

Mutagenesis and toxicity

ScienciaScripts

Imprint

Cover image: www.ingimage.com

This book is a translation from the original published under ISBN 978-3-330-99741-7.

Publisher:
Sciencia Scripts
is a trademark of
Dodo Books Indian Ocean Ltd. and OmniScriptum S.R.L publishing group

120 High Road, East Finchley, London, N2 9ED, United Kingdom
Str. Armeneasca 28/1, office 1, Chisinau MD-2012, Republic of Moldova, Europe
Managing Directors: Ieva Konstantinova, Victoria Ursu
info@omniscriptum.com

Printed at: see last page
ISBN: 978-620-8-61526-0

Index

CHAPTER 1

INTRODUCTION

Pesticides

There have been various names, such as plant remedies, poisons, pesticides, pesticides, agricultural defences, agrotoxics. These are the various names used for chemical substances used to control pests (animal and plant) and plant diseases (Fundacentro, 1998). They are used in pastures, native forests, water, industrial and urban environments, and to combat disease vectors in cities. For around 60 years, the use of agrochemicals has been growing considerably in agriculture, as well as being used in other areas such as wood treatment, construction and public health to treat disease vectors such as dengue fever (Silva et al, 2005).

Living beings are exposed to small quantities of a variety of chemical compounds found simultaneously in the environment. Many of these compounds are included in the class of pesticides, and exposure of humans is most commonly to low concentrations (OLGUN et al., 2004). It is extremely important to analyse the cytotoxicity and genotoxicity of the pesticides most commonly used in fields in order to assess the risk to exposed organisms (DAS; SHAIK; JAMIL, 2007).

Some pesticides have the ability to interact chemically combined in mixtures, in some cases the metabolism of one ends up interfering with the metabolism of the other (BELDEN; LIDY, 2000). These interactions between chemical compounds can often lead to different effects on living organisms. The interaction of pesticides can lead to additive, synergistic or antagonistic effects. These interactions occur during absorption, distribution, metabolism, or at the site of toxic action and excretion of the compound (HODGSON, 1999).

The additive effect requires independent mechanisms of action, so that the action of one toxicant does not interfere with the other and can mask the action of another toxicant in the mixture. Synergistic occurs when different substances interact at the same point in a molecule, enzyme or receptor (HODGSON, 1999).

With the frequent exposure of human beings to complex mixtures of pesticides, it is very difficult to specify genotoxic damage to a particular compound at (GARAJVRHOVAC; ZELJEIC, 2002; BOLOGNESI, 2003). Studies have been carried out in vitro, on different organisms and various cell types, evaluating the genotoxic effect of different classes of pesticides, alone and in combination, allowing the studies to infer specific actions for each compound and understand the actions in humans (AMBROSIO JB et al., 2012).

Exposure to pesticide mixtures occurs accidentally or occupationally (DEMSIA et al., 2007). Occupational exposure to pesticides involves complex mixtures of different types of chemicals, active ingredients, formulation products such as impurities and solvents. As with occupational exposure, environmental exposure of non-target organisms to pesticides occurs through contact with pesticide mixtures, which contain a significant amount of genotoxic compounds (BOLOGNESI, 2003).

Since classical antiquity, farmers have tried to deal with certain problems in their crops, such as fungi, insects, plants and others that harmed production. Roman and Greek records mention the use of arsenic and sulphur to control early agricultural insects. In the United States and Europe, the use of organic substances such as nicotine and pyrethrum extracted from plants was recorded from the 16th century onwards (Silva et al, 2005).

Figura 1. Application of agrochemicals in a monoculture, carried out by a worker wearing PPE (Personal Protective Equipment).

Source: envolverde

One of humanity's greatest challenges is undoubtedly to produce food for an ever-expanding population. However, arable areas are competing for space with urban and industrial areas, making it necessary to return production only to the areas already destined for agriculture, areas that may be worn out, thus making it necessary to use a large quantity of agrotoxins in the search for greater productivity (GRISOLIA, CK, 2005).

This effort to combat the pests that devastated crops began in the middle of the

20th century, with the promise of entrepreneurs and researchers from developed countries committed to increasing productivity and thus reducing or even ending hunger in developing countries. So began the Green Revolution, which was based on a rational production model that adopted the use of hybrid seeds, industrial inputs (fertilisers and pesticides), mechanisation of production, the use of technologies in planting, irrigation and harvesting, as well as management (Moreira, 2000). With the end of major wars, the arms industry found a great solution to maintain profits and exploitation: explosive materials became synthetic and nitrogen fertilisers, deadly gases became agrochemicals and battle tanks became tractors (Fideles, 2006).

Figura 2. Applying pesticides to a plantation using an aeroplane

Photo: outraspalavras

The National Agricultural Development Plan (PDNA), launched in Brazil in 1975, encouraged and demanded the use of agrochemicals, offering investments to finance these inputs and expand the synthesis and formulation industry in the country, which grew from 14 factories in 1974 to 73 in 1985, and has continued to grow (Fideles, 2006).

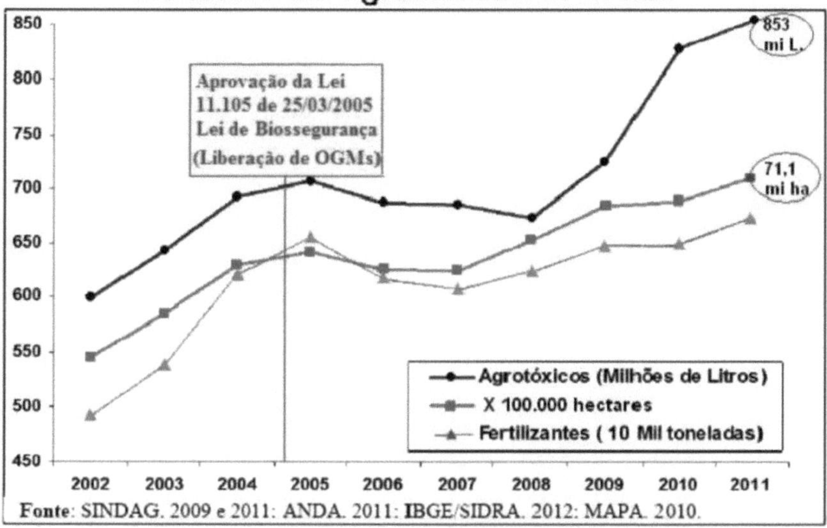

Figure 3: Volume of pesticides used in Brazil between 2002 and 2011. Source: SINDAG; ANDA; IBGE/SIDRA.

Since these incentives to increase the number of these industries, there are currently around 20 major agrochemical manufacturers, making a profit of 20 billion dollars a year and producing around 2.5 tonnes of agrochemicals, 39% of which are herbicides, 33% insecticides, 22% fungicides and 6% other chemical groups. The main players in the market are Syngenta, Bayer, Monsanto, BASF, Down, AgroSciences, Du Pont, MAI and Nufarm. Latin America is a growing and important market in the global context, with sales of agrochemicals increasing sharply between 2006 and 2008 (Sindag, 2009). Brazil has become the world's largest consumer of agrochemicals since 2008. It generated around 6.67 billion dollars in 2008 alone, with a total of 725,600 tonnes, representing 3.7 kg of pesticides per inhabitant. Sales reached around 789,974 tonnes in 2009 (Sindag, 2009).

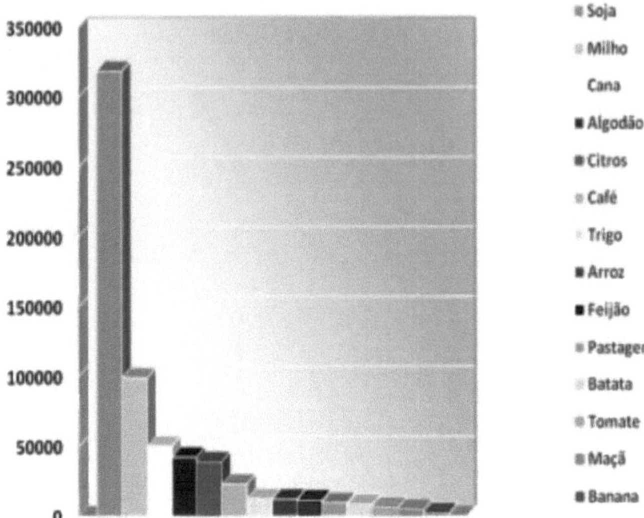

Figure 4: Distribution by crop of the 679,705 tonnes of formulated products consumed in Brazil in 2008.

Source: Sindag, 2008.

In the agricultural sector, the use of pesticides is frequent and in large quantities, especially in monocultures. In the graph above, it can be seen that soya is at the top of the list for pesticide use in Brazil, along with the rapid increase in the area under cultivation: 39% in the South and Southeast and 66% in the Centre-West, in the last 3 years. This is followed by corn and sugar cane plantations, which are also very important for the economy (Rigotto RM, 2013).

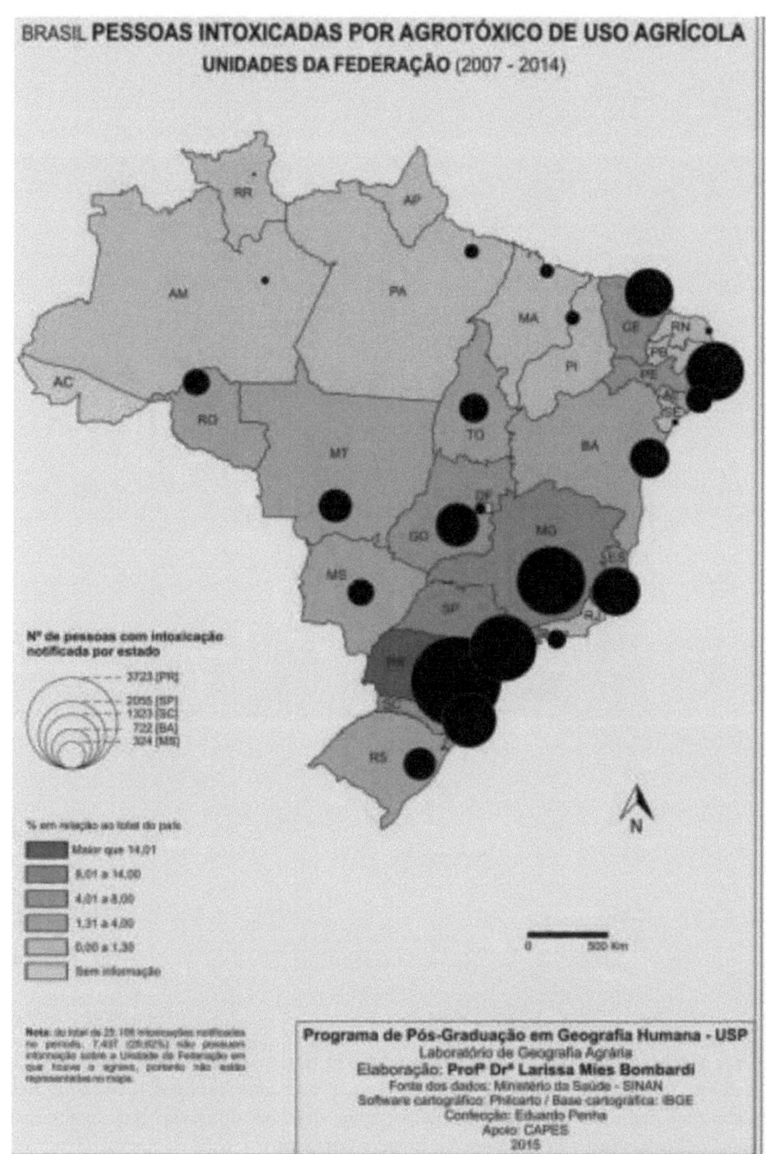

Figura 5. Number of people poisoned by agricultural pesticides in Brazil, broken down by state between 2007 and 2014. Source: Postgraduate Programme in Human Geography - USP. Agrarian Geography Laboratory. Professor Dr Larissa Mies Bombardi.

Chemical substances are sometimes present in high quantities in society, and this can be seen in everyday life, such as in medicines, food, plastics, detergents, paints, pesticides, in industrial and rural environments, as well as in the pollution typical of urban centres, and in a survey carried out in the European Union, , it is estimated that of this range of chemical products, only 7% have some kind of toxicological study (GRISOLIA, CK, 2005).

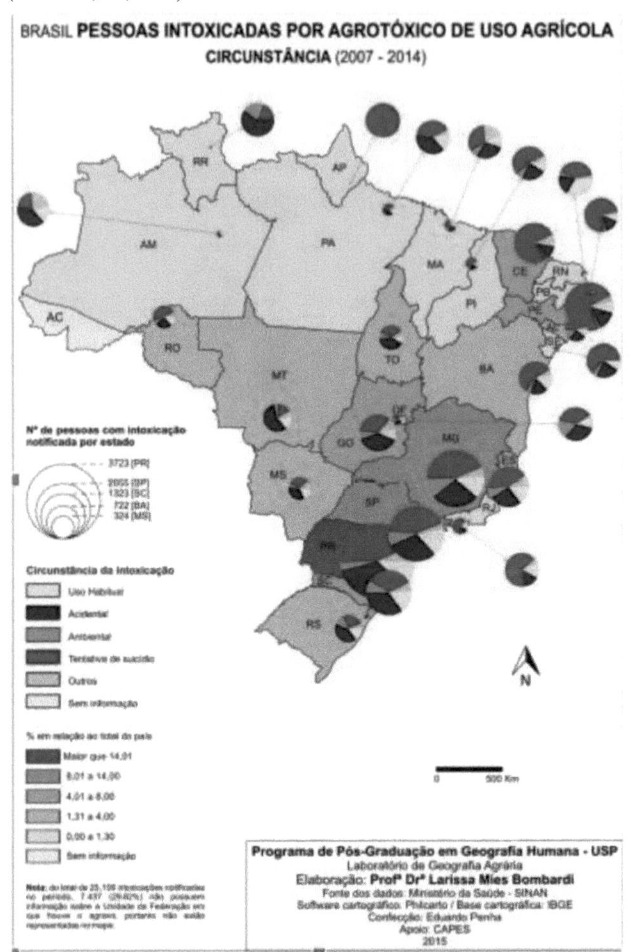

Figura 6. Number of people poisoned by pesticides in Brazil and the circumstances, broken down by state, from 2007 to 2014. Yellow indicates habitual use; blue indicates accidental; green indicates environmental; red indicates attempted suicide; pink indicates other and grey indicates no information. Source: Postgraduate Programme in Human Geography - USP. Agrarian Geography Laboratory. Professor Dr Larissa Mies Bombardi.

Pesticides interfere with physiological life-sustaining mechanisms that are also common to human beings, and can therefore damage health. According to the WHO, pesticides are responsible for around 5 million acute poisonings a year worldwide, mainly in developing countries, which are the biggest consumers (Miranda, 2007). Between 1989 and 2004, 1,055,897 cases of human poisoning by pesticides were reported in Brazil, of which 6,632 died from pesticides (Sinitox, 2004). In 2008, 32.7% of poisonings were caused mainly by agricultural pesticides. The WHO also emphasises the lack of reporting of poisoning cases, with around 50 unreported cases for every reported case (Marinho, 2010).

Class Toxicological	Toxicity	Lethal Dose (50%)	Colour band
I	Extremely toxic	< 5 mg/kg	Red
II	Highly toxic	between 5 and 50 mg/kg	Yellow
III	Medium toxic	between 50 and 500 mg/kg	Blue
IV[7]	Low toxicity	between 500 and 5000 mg/kg	Green

Source: Peres, 2003

Table 1. Classification of pesticides according to toxicity and effects on human health.

An infinite number of compounds with mutagenic potential have been exposed to the human species through its evolution through the ingestion of beverages and foods, which are natural constituents of its diet. More recently, however, through smoke inhalation, occupational activities and, nowadays, a large proportion of synthetic chemical substances, as well as radiation (GRISOLIA, CK ,2005).

With the existing agricultural models, agrochemicals are considered indispensable for agriculture, but on the other hand they are classified as one of the biggest chemical pollutants. North American and European industries are the biggest producers, but developing countries, which are expanding their agricultural frontiers, are the biggest buyers (GRISOLIA, CK, 2005).

Several of the ingredients used by exporting countries are known to be highly toxic to human health and the environment, but have not been banned due to the large number of exports to developing countries (GRISOLIA, CK, 2005).

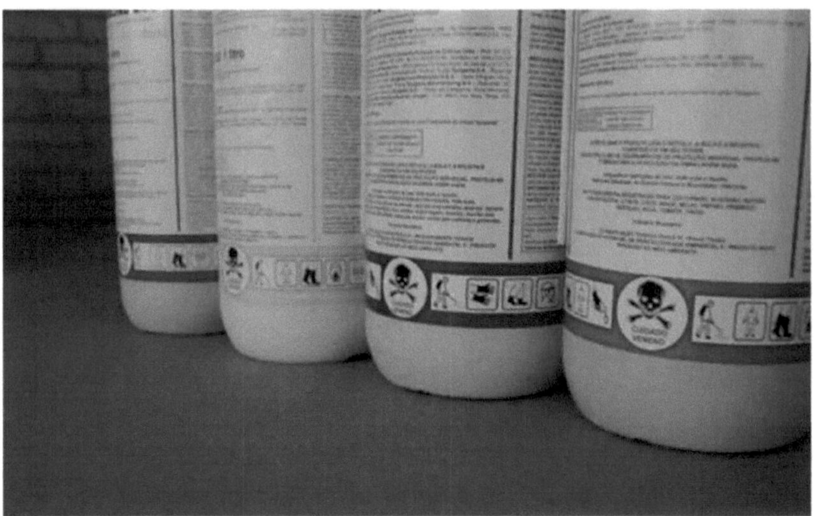

Figure 6 - Pesticide packaging with the toxicity level indicated, red being extremely toxic and green being slightly toxic, yellow and blue being highly and moderately toxic, respectively. Source: aquinoticias.

Occupational toxicology is the area of toxicology that deals with the study of the actions and harmful effects of chemical substances used in the work environment on the human organism, (DELLA ROSA et al., 2003). A large increase in the number of industrial, agricultural and commercial chemical compounds in the aquatic environment leads to various deleterious effects on aquatic organisms (LIVINGSTONE, D.R, 2001).

In Brazil, it is estimated that two thirds of the population is exposed to the harmful effects of these chemical agents, and this exposure can occur through occupation (PERES; ROZEMBERG; LUCCA, 2005), eating and drinking contaminated water (FENSKE, 1997), as well as through the air and dust that reaches their homes (HOPPIN et al., 2006).

There are three ways in which pesticides contaminate the soil (volatisation, leaching and surface run-off), and around 50% of the dose used can remain in the soil when applied to monocultures. Leaching is identified as the main form of groundwater contamination, with rainfall removing up to 2% of the dose of pesticides applied (SIGRH, 2005). These residues reach the environment and can be harmful to humans, potentially altering DNA (EVANS, 1985). Pesticides are generally considered to be chemicals with mutagenic potential, because they have the ability to induce mutations in genetic material, and although they act against certain organisms, there is no absolute selectivity, so they also affect other organisms that are not targeted (BOLOGNESI; MORASSO, 2000).

The figure below represents an analysis carried out by Anvisa of the use of substances (pesticides) that are considered illegal in Brazil, which are present in various types of food, in some cases in high doses.

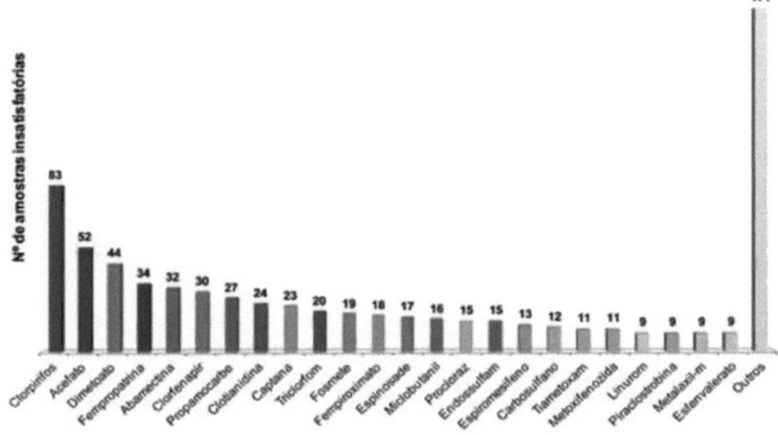

Figura 7. Analysis of illegal pesticides found in inspected foods (Source: ANVISA 2012).

At low concentrations, these substances may have no detectable acute effects on organisms, but they can reduce their survival or damage their health via long-term effects. Such effects can be manifested by greater or lesser damage to somatic or germ cells, and through the development of disorders, such as cancer, which requires long, latent periods before becoming clinically visible (WHITE; RASMUSSEN,1998).

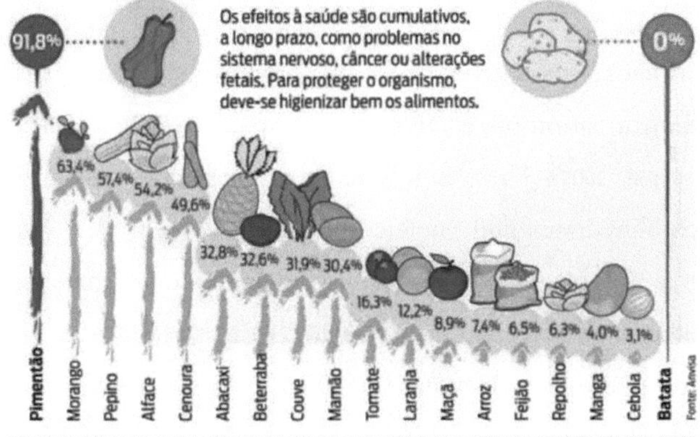

Figure 8. Foods that contain the most pesticide residues. Source: Anvisa

Food can accumulate pesticide residues, which can be harmful to human health over the years when consumed. In Brazil, the Ministry of Health, through ANVISA, monitors around 20 foods and around 234 ingredients. It was observed that 29% of foods in 2009 had unsatisfactory results, either because they were above the maximum permitted limit (>LMR), or because they had residues of unauthorised pesticides that were not suitable for cultivation (NA), or for both reasons (>LMR and NA) (Rigotto RM, 2010).

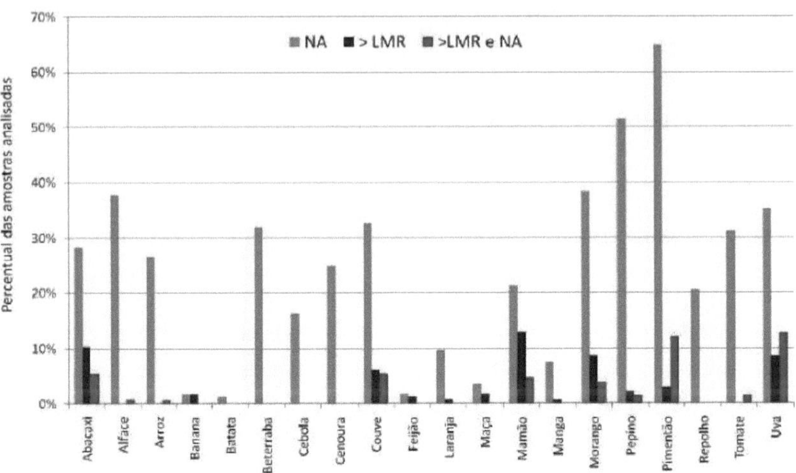

Figure 9. Results of analyses of pesticides in food. Source: Food Pesticide Residue Analysis Programme - PARA.

Exposure to pesticides can cause oxidative stress in organisms, since these contaminants can stimulate the formation of reactive oxygen species (ROS) or alter antioxidant defences (Monserrat et al., 2007). These highly reactive substances can cause damage to lipids, proteins, carbohydrates and nucleic acids (Sevgiler et al., 2004). The antioxidant system comprises a group of low molecular mass antioxidant enzymes, such as ascorbic acid and other non-protein thiols (Winston and Di Giulio, 1991). The antioxidant enzymes include superoxide dismutase (SOD), catalase (CAT) and glutathione reductase (GR), glutathione peroxidase (GPx) and glutathione-S-transferases (GST), which have the functions of detoxifying enzymes that catalyse the conjugation of GSH with a variety of electrophilic compounds (Tuomi et al, 2001).

Pesticides can damage health with chronic effects, such as chromosomal alterations: organophosphate and carbamate insecticides; congenital malformations: fentalamide fungicides, phenoxyacetic herbicides; male infertility: dibromochloropropane nematicides;

Cancer: fungicides, dithiocarbamates, dinitrophenol and pentachlorophenol herbicides; Neurotoxicity: organophosphates and organochlorines; Liver diseases: organochlorines, dipyridyl herbicides; Respiratory diseases: synthetic pyrethroid insecticides, dithiocarbamates; Kidney diseases: organochlorines; Dermatological diseases: organophosphates, carbamates, dioiridyls (FRANCO NETO, 1998; KOIFMAN et al., 2002; PERES et al., 2003; QUEIROZ and WAISSMANN, 2006).

In addition to problems related to human health, serious environmental contamination problems have been observed in monoculture regions, such as contamination of groundwater, in cases such as the Guarani and Jandaíra Aquifers in the states of Ceara and Rio Grande do Norte (COGERH, 2009). Water from dams, ponds, rivers and even water supplied to communities, with up to 12 active ingredients being found in a single sample, similarly contaminating soil, air and neighbouring regions (RIGOTTO RM et al., 2012).

With the intense and widespread use of pesticides in Brazil, it can be considered that most of the population is exposed to pesticides in some way. Workers are the ones who come into contact with pesticides the most, whether in agribusiness companies, family or peasant farming (where the Green Revolution also operates), chemical factories or public health campaigns where they use these products. Another group that can be affected by these products are the communities that live around these agricultural or industrial enterprises, commonly inhabited by the families of the workers of these companies, who live in the so-called sacrifice zones, and there is also the classification of a third group, which are the consumers of the food, which includes practically the entire population (RIGOTTO RM, 2012).

Tebuconazole

Tebuconazole is a systemic fungicide from the triazole chemical group, characterised by its mechanism of action called IBE (inhibitor of ergosterol biosynthesis). It has a preventive and curative action on biological targets that cause serious damage to the production of garlic, oats, potatoes, onions, barley, soya and wheat (Registration: Ministry of Agriculture, Livestock and Supply (MAP) No. 02606). In 2002, the FAO (Food and Agriculture Organisation) issued a global warning on the use of pesticides, in accordance with the International Code of Conduct on the Distribution and Use of Pesticides (FAO, 2002).

Tebuconazole is a very effective fungicide against mould and rust. It belongs to the group of triazole fungicides and acts as an inhibitor of (CYP 51) lanosterol 14 alpha demethylase. In addition to inhibiting the development of fungi, tebucoazole also has the potential to interact with the mammalian cytochrome p450 complex.

The fungicide tebuconazole belongs to the azoles group and is widely used in rice fields. Tebuconazole is classified as toxic to aquatic organisms and can cause long-term adverse effects in the aquatic environment (Bayer CropScience Limited, 2005). The toxic effects of pesticides have been studied in various fish species (MONTEIRO et al., 2006, FERREIRA et al., 2010, TONI et al., 2010).

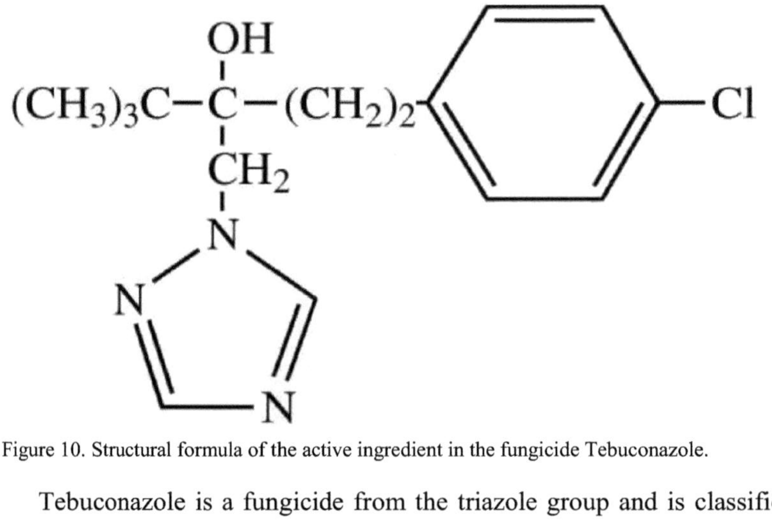

Figure 10. Structural formula of the active ingredient in the fungicide Tebuconazole.

Tebuconazole is a fungicide from the triazole group and is classified by the US EPA as group C- Possibly carcinogenic to humans (US EPA, 2006).Tebuconazole-based fungicide can cause chromosomal aberrations and exchanges between sister chromatids in bovine peripheral blood lymphocytes. Chromosomal aberrations are generally considered to be biomarkers of levels of genetic damage. Both the clastogenic effect (chromosome breakage) and the aneugenic effect (chromosome delays and advances during chromosome segregation) are detected by the micronucleus test. Micronuclei are formed by both acentric chromosome fragments and whole chromosomes left behind during the cell division process (SIVICOVÁ et al., 2013).

Mutagens and genotoxins

De Robertis et al. (2003) and Bernardi (2011) reported that mutations can occur in any cell in the body and can be divided into three categories: genomic, chromosomal or genetic. Gene mutations occur when there are changes involving one or a few nucleotides, while chromosomal mutations are changes that affect the karyotype and can be structural or numerical. Depending on their origin, mutations can be spontaneous, when they occur naturally, or induced, resulting from exposure to agents known as mutagens (GRIFFITHS et al., 2001). The effects caused by chemical, physical and biological agents that can alter cellular components can be classified as teratogenic, carcinogenic and mutagenic (NUNES, 2000).

Genotoxic agents are compounds that have the potential to cause damage to the DNA of organisms, including such effects as DNA breakage, nucleotide modifications, point mutations and chromosomal aberrations (FRENZILLI et al., 2004). These effects mostly occur in somatic cells, altering biological organisation at a molecular level (THEODORAKIS et al., 1998). However, when these alterations affect reproductive cells, they can be passed on to future generations, jeopardising them (LYONS et al., 1997; THEODORAKIS et al., 1998).

Modifications to DNA structure induced by environmental chemical agents can damage vital processes in cells, such as DNA molecule duplication and gene transcription, enabling cell death and the development of tumours (COSTA; MENK, 2000). Mutagens can accelerate and/or increase mutation rates (RIBEIRO; MARQUES, 2003). Rabello-Gay et al. (1991) report that every mutagenic agent can be a potential carcinogen, as unrepaired errors in the DNA possibly become mutations. In addition, the cellular effects caused by exogenous agents can be cytotoxic and genotoxic. Cytotoxic agents can cause a decrease or delay in cell division due to changes in the mitotic apparatus or alterations in the synthesis of macromolecules essential to cell life (MOREIRA et al., 2004). The occurrence of problems arising from genotoxic agents depends on a chain of events involving the quantity and route of entry of the absorbed compound, uptake by the target cells, when these are direct-acting, reaching the nucleus and interacting with the DNA and the fixation of these genetic lesions (RAMEL et al., 1986). Ribeiro et al. (2003) found an increase in mutation rates due to the constant use of products such as drugs, agrochemicals and cosmetics, among others.

In line with the need to quantify the risk of inducing DNA damage and its possible hereditary transmission, various in vivo and in vitro tests have been developed with the aim of identifying and assessing the mutagenic capacity of chemical substances (OGA, 1996). Salvadori et al. (2003) emphasise the importance of detailed, meticulous and orderly studies of techniques to identify and evaluate potential mutagens. Various short-duration tests that evaluate gene mutations, chromosomal damage or DNA lesions are available to assess genetic damage. Therefore, these tests should guide the system for evaluating new chemical agents (MARQUES; RIBEIRO, 2003). Biological tests that evaluate genotoxic activities have been improved in recent years in order to guarantee the reliability of their results (TAKAHASHI, 2003). Mutagenic compounds can cause genotoxic effects in exposed individuals and/or populations, since they are transferred and accumulated through trophic chains. Mutagenic compounds are widely distributed in ecosystems (UMBUZEIRO; ROUBICEK, 2006). Tests that assess genotoxicity, such as the comet assay, and mutagenicity, such as the micronucleus test in vivo or in vitro, are essential in assessing the safety of substance use.

According to Aquino (2010), these tests have the capacity to detect compounds that can induce genetic damage, and he emphasises the importance of combining in vivo and in vitro tests of the same substance in order to verify its possible effects on these systems.

As it is a simple technique, detecting both clastogenic agents that break chromosomes and aneugenic agents that induce aneuploidy or abnormal chromosome segregation, it is one of the most widely used in vivo genetic toxicity tests for assessing the risk associated with exposure to chemical agents (FENECH, 2000; RIBEIRO, 2003).

The genetic material of most living beings, despite being inside cells, protected by membranes, immersed in the cytoplasm and with repair machinery, is not free from constant alterations, i.e. mutations. Mutations, to which genetic material is subject, can be caused by errors that occur during the process of cell division (GRIFFITHS et al., 2006).

Genotoxicity is a broad term which, according to ECHA (2008), refers to processes that lead to alterations in the structure, information content or segregation of DNA. Living organisms are often exposed to genotoxic substances that can cause cellular damage and mutations, but are not necessarily associated with mutagenicity.

While DNA damage caused by genotoxins can be corrected, mutations are alterations in genetic material that cannot be repaired and are passed on to daughter cells. When genetic mutations affect somatic cells, the effects are seen in the individual themselves, as it is possible to observe a reduction or loss of function in the affected cell, which can lead to the development of neoplasms. When the genetic mutation occurs in germ cells, the effects of these alterations are seen in the offspring, which can have anything from organic dysfunctions to the death of the individual (GRIFFITHS et al., 2006). According to LEWIN (2001), these alterations can result from normal cellular processes (spontaneous mutations) or due to the organism's exposure to chemical, physical or biological agents (induced mutations).

Agents that induce DNA damage can affect vital processes such as gene duplication and transcription, as well as chromosomal alterations, leading to the appearance of neoplasms and cell death. Because they induce lesions in the genetic material, these substances are classified as genotoxic (COSTA; MENK, 2000; LEWIN et al., 2001).

Genotoxicity tests provide an indication of induced DNA damage (non-direct evidence of mutation) through effects such as sister chromatid exchange, DNA strand breakage, DNA adduct formation or mitotic recombination.

Several advances in genetic toxicology have expanded the methodologies used to assess the impact of genotoxic agents on organisms (LYONS et al., 1997; BICKHAM et al., 2000;

JHA, 2004).

Micronucleus test (MN)

Until some time ago, the chromosomal aberration test on human lymphocytes and cell lines was the most widely used. Although the micronucleus test was already standardised for rodent bone marrow cells in vivo, its application to human lymphocytes for monitoring populations presented two problems: the fragility of the cells when making the slides, which was solved by modifying the hypotonisation and processing, preserving the cytoplasm and consequently the retention of the micronucleus (MN) inside the cell (ISKANDAR.,1979); (HOGSTED.,1984). The second and main reason was the variability of lymphocyte responses to mitogenic stimuli and the presence of cells in the culture that had not undergone cell division, thus not giving rise to micronuclei, which was reflected in the analysis. The use of radioactive markers was one of the tools used to identify cells that had undergone DNA synthesis. However, it was the discovery of CARTER (1967) who realised in mice that their cells had their cytokinesis interrupted by cytochalasin B, without blocking mitosis, and started using this compound to mark cells that had gone through this division cycle, thus reducing the limitation of the MN test (FENECH; MORLEY,1985).

Micronuclei consist of small amounts of DNA that appear in the cytoplasm when some fragments of chromosomes, chromatids or whole chromosomes are not incorporated into the nucleus of the daughter cell, this occurs in mitosis (DIETZ, J., 2000). Micronuclei are generally 1/5 to 1/20 the size of the nucleus, and there is usually one micronucleus per cell (DIETZ, J., 2000).

The micronucleus test makes it possible to identify an increase in the frequency of mutations in cells that are exposed to a wide range of genotoxic agents (CARVALHO, MB et. al., 2002). The micronucleus test is considered by laboratories to be a cheap, fast, non-invasive test that can be repeated several times (CARVALHO, MB et. al., 2002).

The Micronucleus (MN) test is considered by many authors to be one of the most promising techniques for evaluating mutagenic effects induced by chemical agents

(MATSUMOTO et al., 2006). Genotoxic agents are those that interact with DNA, altering its structure or function. When these alterations become fixed and acquire the capacity to be transmitted, they are called mutations (UMBUZEIRO, GA et. al., 2003).

The micronucleus test and nuclear anomalies indicated the probable presence of agents that are compromising the quality of the water in the lakes tested, suggesting that chronic exposure to these waters could affect the biota present in these ecosystems, as well as the health of the local population (MIRANDA, CT, 2009).

Allium cepa

The Allium cepa method is an effective test for assessing chromosomal alterations and is validated by the International Programme on Chemical Safety (IPCS-WHO) and the United Nations Environmental Programme (UNEP) as a good test for in situ analysis and monitoring of the genotoxicity of environmental substances (CABRERA, GL; RODRIGUEZ, DMG, 1999).

Figure 11. Allium cepa test, bulbs with the roots immersed in water to evaluate compounds. Source: ciência hoje.

The Allium cepa test is commonly used to analyse various substances to determine their mutagenic potential. It is estimated by the frequency of chromosomal aberrations and breaks, indicating the risk of aneuploidy and providing valuable information on the relationship between the evaluation of environmental samples (RANK; NIELSEN,1993).

The mitotic index and replication index are parameters used to analyse appropriate cell proliferation, which can be measured using the Allium cepa plant test system (GADANO et. al., 2002).

The efficiency of the Allium cepa system is due to its kinetic proliferation characteristics, the rapid growth of its roots, the large number of dividing cells, its high tolerance to different growing conditions, its availability throughout the year, its easy handling and the fact that it has a small number of chromosomes (2n=16) (FISKEJÓ, 1994).

CHAPTER 2

OBJECTIVES

General objective

The general objective of this study was to evaluate the fungicide Tebuconazole and its possible mutagenicity.

Specific objectives

- Evaluating the mutagenicity of Tebuconazole in meristematic cells of Allium cepa
- To evaluate the mutagenicity of Tebuconazole by performing the micronucleus test on human peripheral blood cells.

CHAPTER 3

MATERIALS AND METHODS

Allium cepa

The test solutions were obtained from the determination of the $CL_{50/48}$ for *Poecilia reticulata* (17.57µL.L^{-1}) and serial dilutions were made to obtain concentrations of 13.18µL.L^{-1}; 8.79µL.$L^{(-1)}$) and 4.39µL.$L^{(-1)}$. The *A. cepa* bulbs were subjected to root growth in the test solutions for three days after the initial growth period in distilled water for three days, and the root samples were collected and fixed in Carnoy. The slides of the meristematic tissue were obtained using the *squash* method according to the protocol of FISKEJÓ (1985). The slides were made by removing the root coif and placing it in a microtube with 1N HCl for 10 to 15 minutes, then placing it in a microtube with acetic orcein for 10 minutes and then placing the coif on a slide, dripping a drop of acetic acid and squashing it. To analyse the slides, white light microscopy was used with a 40x objective, making it possible to identify migration errors (adhesions, delays, advances and chromosome bridges) in the metaphase and anaphase phases of cell division.

Micronucleus

For the micronucleus test, 10 ml of human blood was taken (from a healthy, non-smoking, non-alcoholic person who is not taking medication), and then a blood culture was carried out at 6 concentrations (10µL.$L^{-1)}$ 100%; (7.5 µL.L^{-1}) 75%; (5.0 µL.L^{-1}) 50%; (2.5 µL.L^{-}ı) 25% ; dilutions obtained from Cl 50 of 13.5 µL/ml; and the negative control with culture alone and positive with the addition of MMC (Methyl methane sulphonate) 10µL at a concentration of 5 µL/ml.The culture was carried out in tubes with 4ml of RPMI, and 1ml of blood was added, 1 of foetal bovine serum, 100 µL of phytohaemagglutenin, 100 µL of cholscicin, the culture was carried out initially without the addition of the test solutions and after 24h the concentrations of the test substance (Tebuconazole) were added.After 48 hours, the culture was interrupted and cleaned with Carnoy's solution, so that the slides were made using the drip technique and stained with panothecin dye. 1000 cells were examined per slide. Only cells with an intact cytoplasmic membrane were considered. Micronuclei must not exceed more than 1/3 of the size of the nucleus and not less than 1/30. MNs must be clearly separated, with distinguishable edges, not connected to the main nucleus, of the same colour and refringency.

CHAPTER 4

RESULTS AND DISCUSSION

Cytotoxicity, genotoxicity and mutagenicity assessments carried out on plants are considered to be effective and simple tests, capable of providing excellent results when compared to other test systems (GRANT, 1982; CHAUHAN et al., 1999). As direct recipients of pesticides, plant species such as Allium cepa, Arabidopsis thaliana, Hordeum vulgare, Tradescantia sp, are frequently used to assess the toxic effects of various compounds in the environment (MA et al., 1995).

The results obtained from the determination of the Cl50 of the fungicide Tebuconazole carried out on Poecilia reticulata over a period of 96 hours was 17.57µl/L.

As the fungicide Tebuconazole belongs to the class of extremely toxic agrochemicals, its Chl 50 was 17.57µl/L in *Poecilia reticulata* for 96h. Compared to the herbicide Roundup, and to other test organisms, the Chl 50 of young tilapia (*Oreochromis niloticus*) was found to be 17.5 µl/L in 24h and 96h periods, and for adult tilapia a result of 46.9 and 36.8 was found in 24h and 96h (JIRAUNGKOORSKUL *et al.,* 2002), showing a similarity with Tebuconazole in the Chl of young tilapia.

Molecular studies have been carried out to assess the mutagenicity of pesticides, using the RAPD marker. The insecticide dimethoate was assessed in zebrafish, where significant differences were reported between treated and control samples (ZHIYI & HAOWEN, 2004). In a study with paranitrophenol, a greater significance was observed between the treated and control samples as the concentrations increased, using the RAPD marker (ENAN, 2007). And in a study with barley seedlings, a disappearance of DNA bands was observed with increasing cadmium concentrations (LIU *et al.*, 2005).

Losses and gains of bands have been observed using RAPD in fish exposed to the active ingredient carbofuran (MOHANTY *et al.*, 2009). Using ISSR and RAPD markers, the effects of the herbicides atrazine and roundup, such as changes in band numbers, intensity and position, were observed in *Biomphalaria alexandrina snails* (BARKY *et al.*, 2012).

Another study analysing Cl 50 of the active ingredient in Roundup on carp (Cyprinus carpio) found a value of 620 $mg.L^{-1}$ over a period of 96 hours, but the authors observed that sublethal concentrations of less than 1% of Cl 50 caused damage to the carp's gills (NESKOVIC et al. 1996).

The in vitro animal test and the in vivo Allium cepa test to assess cytotoxicity have been validated by several researchers, as the results obtained are very similar (TEIXEIRA et. al., 2003; VICENTINI et. al., 2001).

Plant metabolism is different from animal metabolism, but the Allium cepa plant test system is an excellent parameter for cytotoxic analysis, and has been used to prevent human health by avoiding the consumption of various products, through observations of chromosomal alterations in the Allium cepa test (FISKEJÓ., 1994).

It is worth emphasising that studies on Allium cepa are always designed to demonstrate toxicity. Changes in root growth indicate phytotoxicity and in the mitotic index indicate cytotoxicity (OLIVEIRA et al., 2011); chromosomal aberrations are capable of assessing genotoxic and cytotoxic effects, while nuclear aberrations assess cytotoxic effects (FISKEJO, 1985; LEME, MARIN- MORALES, 2009).

In the Allium cepa test, three slides were counted per treatment, a total of 9 slides per concentration, and each slide counted 100 anaphases and considered normal and abnormal, and 100 metaphases considered normal and abnormal.

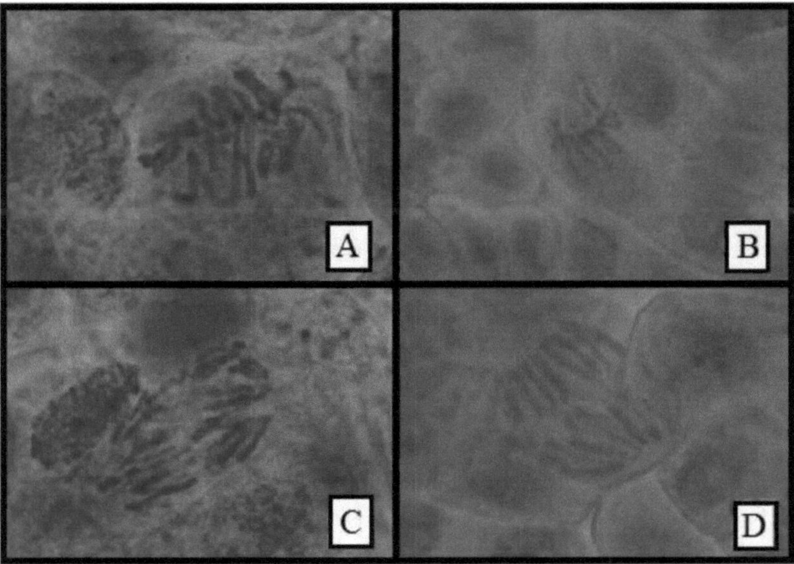

Figure 12. A) Abnormal metaphase; B) Normal metaphase; C) Abnormal anaphase; D) Normal anaphase; at 100x magnification. Source: Replicon Research Centre - PUC GO.

In the Allium cepa test, the highest mean was the concentration of 13.18µl.L^{-1} , but there was no correlation between the means of the concentrations of 17.57 µl.L^{-1}, 8.79 µl.L^{-1} and 4.39 µl.L^{-1} when compared to each other, so there was no significant difference between the comparison between the exposed group, but there was a significant difference between the mean of the exposed group when compared to the negative control (Figure 13).

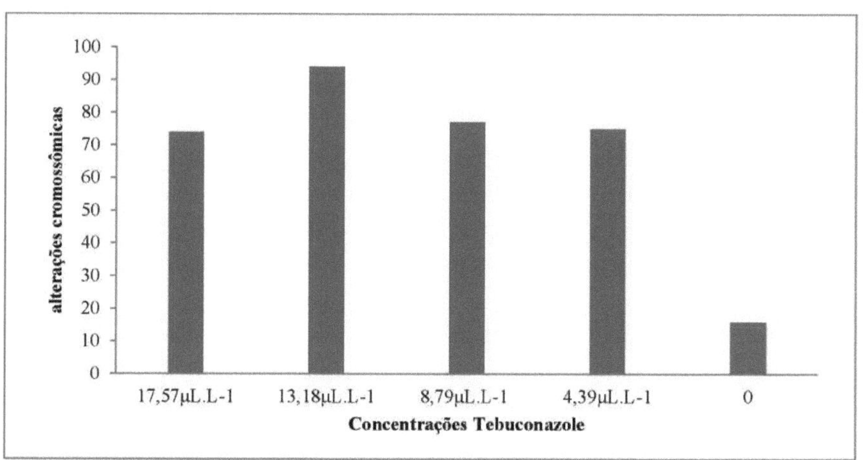

Figure 13. Average chromosomal alterations between concentrations of the fungicide Tebuconazole in the Allium cepa test.

The Aliium cepa test system used linear correlation models, Spearman's correlation, Tukey's test and ANOVA to carry out the statistics. The p-value obtained was p=0.0000019, showing a significant difference.

The concentration with the highest average, 13.18 µl.L^{-1}, is not the highest concentration, the concentration of 17.57 µl.L^{-1} (100%) was the one that showed the lowest average, but as there was a significant difference between the concentrations of Tebuconazole and the negative control, the fact that the 100% concentration was the lowest average may have been due to the toxicity of the fungicide having influenced the cell division cycle, being an inhibitor, or due to a difference in the illumination of the bulbs in root growth, among other reasons, but not influencing the results of the mutagenic and cytotoxic potential that Tebuconazole showed in the Allium cepa test. The fungicide Tebuconazole in the plant test system (Allium cepa), showed a significant difference when the averages of the exposed group were compared with the negative control, and it was observed that there was a greater number of alterations at the 75% concentration. There was no dose-dependent correlation, but possibly if the concentration were higher there would be a higher average number of alterations.

Studies have reported differences between active ingredients and commercial formulations, as surfactants and so-called inert components can increase the toxicity of these formulations (BENDER et al., 2006). One study found that surfactants and inert components can increase the total toxicity of pesticide formulations by up to 50 per cent (NONDILLO et al., 2007).

A study carried out with Roundup, containing glyphosate, evaluated some species of fish, where genotoxicity of this agrochemical was observed (BRAGUINI, 2005). Another study assessed glyphosate, which is present in the Roundup formulation, using the Allium cepa test system, where root growth inhibition, anaphase-telophase abnormalities and micronuclei were observed, showing the genotoxicity of this compound (KRUGER et al., 2009). In the study by Kruger et al. (2009), the genotoxicity of the compound beta-cyfluthrin, used in pesticides, was demonstrated. In a study in which pyrethroids were tested, cyfluthrin showed genotoxicity both in vitro and in vivo (TISCH et al., 2005; ELIK et al., 2005). However, cyfluthrin did not show dose-dependent genotoxicity. And the cytotoxic effects of pyrethroids have been reported (CABALLO et al., 1992; FISHEL et al., 2005).

Agrochemicals containing mancozeb have been reported to cause oxidative and genotoxic damage and may be related to diseases, including various types of cancer. Due to its low toxicity and environmental persistence, the use of the fungicide mancozeb has been allowed worldwide, even after problems of chronic exposure to this pesticide were reported. A study carried out on female rats observed sterility, cycle alterations, a decrease in ovarian size, among other things, after exposure to mancozeb (BINDALI & KALIWAL, 2002). Decreased sperm count and increased frequency of sperm with aberrant morphology have also been reported in male rats (KHAN & SINHA, 1996). Mancozeb has also been shown to alter endocrine gland secretion (BISSON & HONTELA, 2002). It also induces neuronal degeneration and carcinogenic potential has been described (JARRARD et al., 2004; BELPOGGI et al., 2002). In a study carried out with mancozeb on allium cepa, the cytotoxic and genotoxic action of this pesticide was found (KRUGER et al., 2009). Genotoxic action was also observed in the lymphocytes of rural workers exposed to mancozeb (BINDALI & KALIWAL, 2002).

A study carried out with the active ingredient in the pesticide Tebuconazole (tebuconazole) found cytotoxicity, genotoxicity, mutagenicity and phytotoxicity effects, with macroscopic, microscopic and molecular analyses carried out using the Allium cepa system, It was observed that the highest concentrations caused the greatest alterations, and the principle tebuconazole, among the principles tested in the study, caused the greatest damage, even at lower concentrations (BERNARDES PM et al., 2014).

Macroscopic tests (germination, root growth) and microscopic tests (mitotic index, chromosomal and nuclear aberrations) are carried out on the root tip to assess the toxic effects of compounds. Effects such as delayed germination are considered phytotoxic effects, and reduced germination frequency is a parameter widely used to assess toxicity (VALERIO et al., 2007). Root growth is involved with an increase in the number of cells and elongation during cell differentiation (HARASHIMA; SCHNITTGER, 2010). Macroscopic data can be directly related to the results of microscopic evaluations, and the increase in the number of cells is linked to mitotic cell division, with successive mitotic cycles being necessary for root growth over a period of time (ANDRADE et al., 2010).

In the study by Bernardes et al. (2014), the dose-dependent effect of the active ingredients of the fungicides was verified, which became evident with the increase in alterations according to the increase in concentrations. It was found that germination and root growth were reduced at the highest concentrations of the active ingredients, with the exception of one, procymidone. The principle tebuconazole showed the greatest toxicity, cytotoxicity and mutagenicity at all the concentrations assessed. It affected seed germination at all concentrations, with the greatest reduction at the highest concentrations, showing cytotoxic and phytotoxic action, preventing cell division and seedling development. It also induced cell death, showing a decrease in the mitotic index, due to the high frequency of nuclear aberrations and DNA degradation by apoptosis.

The mitotic index decreased in relation to the negative control for all the active principles, thus indicating the cytotoxic action of the principles on the roots of Allium cepa, especially the difenoconazole and tebuconazole principles. There was also a higher frequency of nuclear and chromosomal alterations for all the active ingredients compared to the negative control. Tebuconazole showed more chromosomal aberrations at the concentrations evaluated. The other principles evaluated, at or above the recommended concentration, were cytotoxic, showing broken chromosomes, lost chromosomes, bridges, sticky chromosomes, c-metaphase, polyploid c-metaphase (BERNANRDES PM, 2014). This cytotoxicity has also been found in the insecticide carbofuran (SAXENA et al., 2010) and the herbicide atrazine (BOLLE et al., 2004).

Studies have shown that compounds evaluated with the Aliium cepa test, such as cadmium, reduced root growth at higher concentrations (LIU et al., 2009). On the other hand, copper sulphate, which is often used in treatments against pests in crops, caused toxic effects due to delayed root growth at lower concentrations (YILDIZ et al., 2009).

In one study, two pesticides were analysed using the Allium cepa test and checking the frequency of micronuclei, chromosomal aberrations and the mitotic index. After exposure to the insecticide imidacloprid for 24 hours, a slightly higher induction of micronuclei and chromosomal aberrations was observed when compared to the negative control, but it was not significantly different; only one concentration showed significantly higher micronucleus frequencies. What was also observed in the analyses were the results of the mitotic index, which showed no significant changes, showing that the insecticide no has cytotoxic effects on the cells studied. The other pesticide evaluated was the herbicide sulphentrazone, where the highest concentration caused total inhibition of cell division, which made it impossible to analyse genotoxicity and mutagenicity. Cells exposed to the 12 mL/L concentration showed high rates of cell death by apoptosis. At the other concentrations, which were slightly lower, it was possible to carry out cytotoxicity, mutagenicity and genotoxicity tests. The concentration of 2.4 mL/L (the second highest concentration) was able to induce significant rates of chromosomal aberrations in A. cepa meristematic cells, while the lowest concentration (0.12 mL/L) was able to induce significant rates in A. cepa cells (AMBROSIO JB et al., 2012).

Analyses of mitotic indices in the work by Ambrosio et al. (2012) showed that the herbicide sulphentrazone did not have significant cytotoxic effects. The mixture of these two pesticides was analysed, with three concentrations of the mixture, and it was found that the M3 mixture was able to induce significant micronucleus results in the meristematic cells, while the M1 mixture caused a significant decrease in the mitotic index. The study carried out a recovery of the cells after exposure to the pesticides, for the insecticide imidacloprid and for the herbicide sulphentrazone, after 48 hours of recovery, significantly increased chromosomal aberrations were observed for the insecticide at all concentrations compared to the negative control, but no significant results were observed for micronuclei and alterations in the mitotic index, and in the 72-hour recovery, significant indices of chromosomal aberrations were observed. The herbicide sulphentrazone, on the other hand, was found to significantly induce chromosomal aberrations at a concentration of 1.2 mL/L during the 48-hour recovery period, while the concentration of 0.12 mL/L induced higher levels of micronuclei in the meristematic cells of A. cepa, and also induced an increase in the mitotic index. In the 72-hour recovery period, the 0.12 mL/L concentration showed a significant increase in chromosomal aberrations and micronuclei, but a decrease in the mitotic index.

The mixture of the insecticide imidacloprid with the herbicide sulphentrazone, after 48 hours recovery, significantly induced chromosomal aberrations for all three concentrations of the pesticide mixtures (M1; M2; M3). The M2 concentration significantly induced micronuclei in A. cepa cells. The M1 and M2 concentrations induced significant cytotoxic effects, due to the decrease in the mitotic index. Cells were also observed in the process of cell death at the M1 concentration within 48 hours of recovery, and for M3 within 72 hours of recovery (AMBROSIO et al., 2012).

It has been found that glyphosate-based products such as Roundup® are potentially more mutagenic than glyphosate alone. This is due to the presence of surfactants that potentiate the action of glyphosate (COX, 1998). One study showed the mutagenic action of Roundup® when compared to glyphosate alone, with various tests on rats (Swiss CD1) to assess DNA damage and chromosomal effects in vivo and in vitro (BOLOGNESI et al., 1997).

Chromosomal aberrations were observed in Allium cepa cells exposed to Roundup®, but not to glyphosate (RANK et al., 1993). In another study with the plant species Crepis capillaris exposed to various concentrations of Roundup®, no increase in the frequency of micronuclei was observed in root cells, and the concentrations analysed showed no mutagenicity (DIMITROV et al., 2006). In an in vitro study on human and bovine peripheral lymphocytes, chromosomal aberrations were found when exposed to glyphosate (LIOI et al., 1998).

In the work of DISNER, GR et. al (2011), a study was carried out on the herbicide Roundup®, which through the cell growth inhibition test on Allium cepa, the toxicity of the compound was observed, and the higher the concentration of the compound in contact with the onion bulbs, the lower the cell division in the roots. Even very low concentrations can cause some degree of toxicity to the organism. A significant increase in micronuclei was also observed in the cells of Allium cepa roots that were in contact with glyphosate.

In the work of VENTURA, BC et. al. (2004), an evaluation of the herbicide Atrazine was carried out using the Allium cepa test to analyse chromosomal alterations. The test was carried out with 48 hours of exposure and 20 hours of dilution, and it was observed that the concentrations of the exposed group showed a significant difference when compared to the negative control, and there was a significant difference between the concentrations, and this for both treatments. It can be seen that most of the chromosomal alterations are related to the loss of genetic material, and the frequency of aberrations in the (75%) concentration was considered significant, while the (100%) concentration induced a lower number of chromosomal aberrations, which can be explained by the inhibition of the mitotic index that this concentration generated. The study also carried out a period of recovery of the root cells with milliQ water, and it was subsequently observed that at the end of the recovery period there was a reduction in chromosomal aberrations.

Also in the work by VENTURA, BC et al. (2004), a cell death test was carried out on the root meristem, and it was observed that the higher the concentration of the herbicide Atrazine, the greater the cell death. This cell death was observed for all concentrations, but it was more restricted to the procambium and with increasing concentrations it was also observed that the regions of the fundamental meristem and protodermis were also compromised. There was thus greater cell damage at the higher concentrations.

In a study of the herbicide Trifluralin, according to FERNANDES, TCC et. al., (2005), they carried out mitotic index inhibition tests on Allium cepa cells, with a 24-hour treatment, after which there was a 24-hour cell recovery period and then a 48-hour recovery period. The mitotic index inhibition test for the 24-hour exposure showed a significant difference with the control group, since there was a decrease in the mitotic index at concentrations of 50%, 75% and 100%. In the 25% concentration there was no significant observation in any treatment, but it and the 50% concentration had an increase in the mitotic index when the recovery periods occurred, which unlike the other concentrations, there was no increase in the mitotic index. In the same study, chromosomal alterations were analysed in Allium cepa, and a significant difference was also observed between the concentrations of the herbicide Trifluralin and the negative control. Some of the alterations were the result of the sum of the effects occurring in different phases of the cell cycle and/or due to the persistence of the herbicide in various phases of the cell cycle, which may potentiate its effect.

Some studies have investigated the genotoxic and mutagenic effects of the herbicide Imidacloprid on non-target organisms such as human lymphocytes (FENG et al., 2005; DEMSIA et al., 2007), rat bone marrow cells (KARABAY; OGUZ, 2005, DEMSIA et al., 2007), amphibians (FENG et al., 2004), aquatic organisms (JEMEC et al., 2007) and microorganisms (KARABAY; OGUZ, 2005; KREUTZWEISER et al., 2007), obtaining positive and negative results for the genotoxicity and mutagenicity of the compounds evaluated. Another pesticide that has been evaluated is sulphentrazone, where some studies using rats have not registered positive results for genotoxicity and carcinogenicity (EPA, 2006; CASTRO et al., 2007), but have shown positive results for teratogenicity (EPA, 2006).

Various alterations were analysed in all concentrations and all treatments, both in the 24-hour and the recovery treatments. Another analysis showed the presence of micronuclei, except in the 100 per cent concentration, mostly in the interphase and prophase phases.

In the micronucleus test on human T lymphocytes, the Cl 50 used was the same as for Poecilia reticulata, but at a lower dilution because the test systems are different in several respects, so a dose of 13.18 $\mu l.L^{-1}$ was used as the Cl 50 for the dilution and the Tebuconazole dilutions were carried out afterwards.

In the test, 1000 binucleated cells were counted per slide, with three repetitions at each concentration, 3000 cells were counted per treatment and 66 micronuclei (MN) were found. Normal binucleated cells and binucleated cells with micronuclei were analysed.

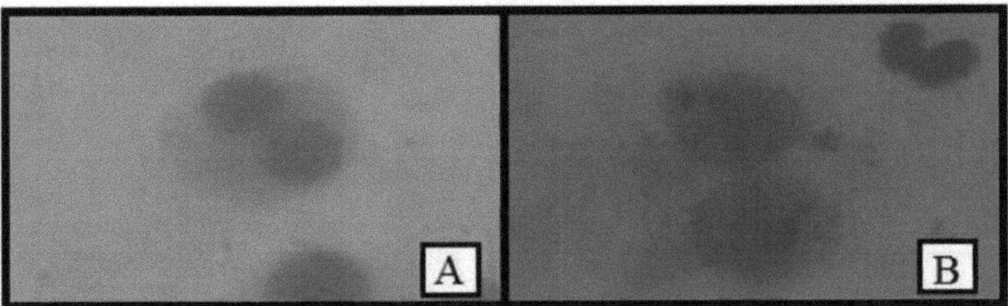

Figure 14. A) Binucleated cell; B) Binucleated cell with micronucleus.
Source: Replicon Research Centre - PUC GO.

Pearson's correlation and the Tukey, Anova and Shapiro-Wilk statistical tests were used for the micronucleus test. A p-value of 0.0008 was obtained, proving to be significant.

The highest concentration of 10 µl/ml showed the highest average number of micronuclei when compared to the other concentrations of 7.5 µl; 5.0 µl; 2.5 µl and when compared to the negative control. The other concentrations of 7.5 µl; 5.0 µl; 2.5 µl also showed significant differences when compared to the negative control.

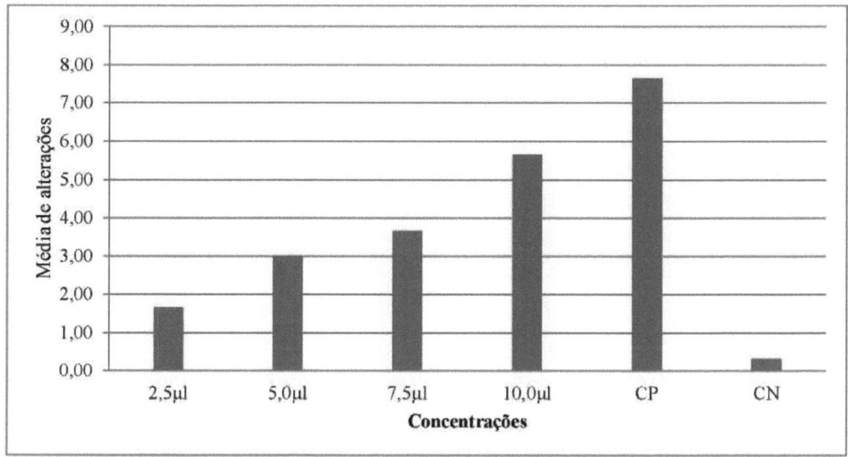

Figure 15. Differences in micronucleus averages between Tebuconazole concentrations and the negative and positive control.

There was also a dose-dependent relationship, as the higher the dose of Tebuconazole, the higher the number of micronuclei found.

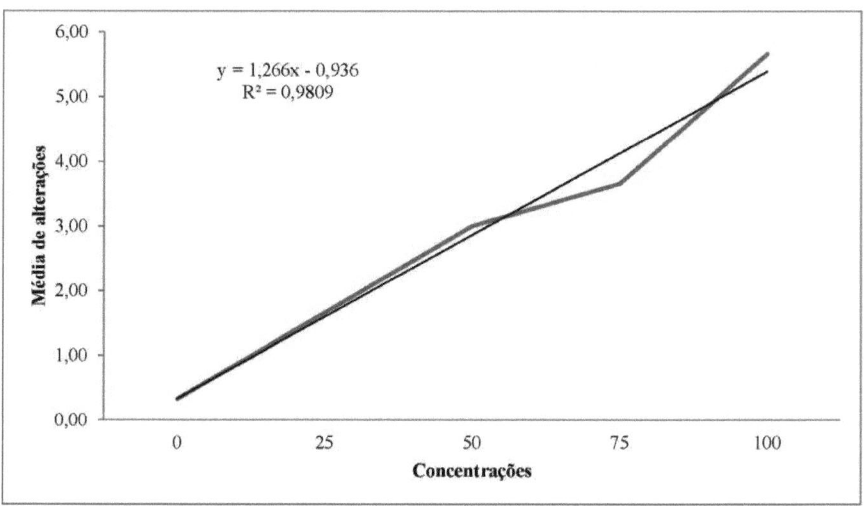

Figure 16. Micronucleus increase ratio according to the increase in Tebuconazole concentration.

There was a correlation between the mean concentrations, as there were significant differences between them, as well as the concentrations of 10μl; 7.5 μl; 5.0 μl; 2.5 μl, which also showed differences compared to the control group.

The study carried out by NEVES, JT (2010) with workers exposed to different types of pesticides found a significant difference when compared to the control group, meaning that the people who work directly with pesticides had a higher number of micronutrients than the people who were used in the control group, which are people who are not directly involved with pesticides. There is a division of micronutrient categories, categorised with greater alliteration. In the control group, most of the individuals are in the lower ciasses, and as the ciasses increase, the number of individuals present decreases, until the last ciase is absent. In the exposed group, the number of more frequent individuals tends to increase as the ciasses increase, but in the last ciasses there is a decline.

Studies have been carried out in vitro to assess the effects caused by mixtures of different pesticides (BIANCHI-SANTAMARIA et al., 1997; SABALIUNAS et al., 1998). In one study, it was found that human iymphocytes had a significant induction of sister chromatid exchanges when the cells were exposed to a mixture of 4 pesticides; on the other hand, no significant difference was observed when analysed separately (DOLARA et al., 1992).

In the study with the insecticide imidacioprid and the herbicide suifentrazone, in human cells, it was noted that the pesticides induced significant mutagenicity for three assessed concentrations of the insecticide, and for two concentrations of the herbicide. However, with mixtures of the two pesticides, no significant frequency of micronuclei was observed. Most of the micronuclei formed by exposure to the insecticide were cytogenic. On the other hand, the micronuclei induced by the herbicide or the mixture with the insecticide showed micronuclei of cyastogenic and aneugenic origin. The comet assay, which evaluates the genotoxicity of the compounds, found genotoxicity for some concentrations of the insecticide, as well as genotoxicity for two mixtures, concentrations M2 and M3. No significant genotoxicity results were found for the herbicide. Overall, the study found that both the insecticide and the herbicide were capable of inducing micronuclei in human cells, indicating mutagenic action, where they also observed, but not significantly, a slight increase in the frequency of micronuclei in the lowest concentrations of the two pesticides and in the combination of the lowest concentrations of the insecticide and the herbicide (AMBROSIO et al., 2012).

Some studies carried out with toxic compounds have shown that higher concentrations of a compound can cause less damage to DNA than lower concentrations, due to the greater activity of antioxidant enzymes and the greater damage potential of the higher concentration (AU et al., 1988; 1990; CAVALLO et al., 2010).

Other studies also involving the insecticide imidacloprid have shown mutagenic, cytotoxic and genotoxic effects on cells in other organisms and cell types, such as the in vitro tests carried out on human lymphocytes and amphibian erythrocytes, which were exposed to different concentrations of the pure insecticide (> 95%) and the commercial insecticide Confidor® (17.8%).95%) and the commercial insecticide Confidor® (17.8%), indicating positive results for both micronucleus induction and induction of primary DNA damage, which were observed by comet assay (FENG et al., 2004, 2005; COSTA et al., 2009).

Studies with the herbicide sulphentrozoan have shown no genotoxic or carcinogenic effects, although others with the same herbicide have shown mutagenic effects, but these are inconclusive (BIGGER; CLARKE, 1992; CALIFORNIA EPA DEPARTMENT OF PESTICIDE REGULATION, 2006). However, many studies have shown negative mutagenesis for the pure herbicide and not for the commercial herbicide, and the diversity of cells used in the studies may interfere with the different results (BIGGER, CLARKE, 1992; PUTNAM, YOUNG, 1992; MURLI, 1996).

In a study carried out by our group, lead author Pena RV et al. (2011) analysed the fungicide Tebuconazole in the erythrocytes of the fish Poecilla reticulata (Guppi), where the concentrations of Tebuconazole were measured and the types of erythrocyte assessed, such as normal erythrocyte, erythrocyte with reniform nucleus, erythrocyte with lobate nucleus and micronucleated erythrocyte, compared to the cells exposed to the negative control. Although changes were observed in the erythrocytes in the exposed cells, they were not characterised as significantly different from the control, showing the differences between the models used.

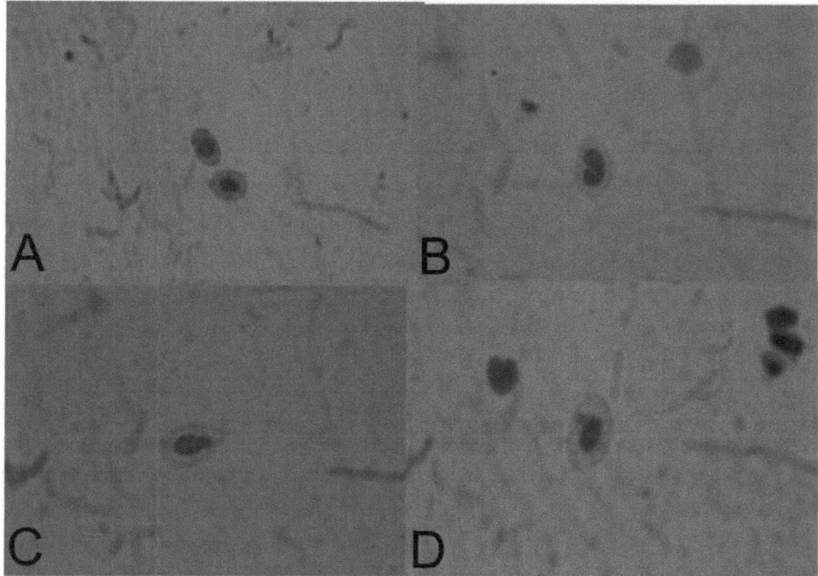

Figure 17. Erythrocytes from P. reticullata exposed to the fungicide Tebuconazole. A) Normal erythrocyte;

B) Erythrocyte with reniform nucleus; C) Erythrocyte with lobed nucleus; D) Micronucleated erythrocyte. Source: Replicon Research Centre - PUC - GO.

Studies based on the genotoxicity of glyphosate and formulations based on this product, such as Roundup®, have shown great variations in the results, and these differences can be attributed to the fact that different formulations were tested, application doses, methodologies employed and model organisms used in the studies (ÇAVAS & KONEN, 2007). According to some authors, glyphosate and formulations of this agrochemical can influence both the absence (WILDEMAN & NAZAR, 1982; GRISOLIA CK, 2002) and incidence of damage (CLEMENTS et al., 1997).

In a study carried out with a mixture of the insecticides lindane (organochlorine), malathion (organophosphate) and permethrin (synthetic pyrethroid), the cytotoxic effect of this mixture was found to induce cell necrosis and apoptosis, with significantly higher results compared to the insecticides tested separately (OLGUN et al., 2004).

The mutagenic effect of glyphosate has been observed in Carassius auratus (goldfish), where the lowest concentration capable of inducing a significant increase in the number of micronuclei in the exposed fish was 5 $mg.L^{-1}$ (ÇAVAS & KONEN, 2007).

CN	3.2$\mu l.L^{-1}$	6.5$\mu l.L^{-1}$	9.8$\mu l.L^{-1}$	12.3$\mu l.L^{-1}$
1	2	4	2	3
2	2	2	2	3
1	2	4	6	4
0	4	4	1	3

Table 2. Values of micronuclei and nuclear erythrocytic alterations found per individual.

In a study of the herbicide Roundup, carried out on the fish species Prochilodus Lineatus, genotoxicity and mutagenicity were analysed using peripheral blood and gill cells, initially using the comet assay, then the micronucleus test and erythrocyte alterations, in the comet assay for erythrocytes exposed to the pesticide, the mean scores at 6 and 96 hours were significantly higher when compared to the negative control, while the mean scores for gill cells showed a significant increase at 6 and 24 hours (CAVALCANTE DGSM et al., 2008).

According to the work of CHAVES, TVS (2007), a study was carried out with rural workers in some municipalities in the state of Piauí, where they are exposed to pesticides in their daily lives, and the micronucleus test was carried out on cells from the oral mucosa. Biological samples were collected from workers exposed to pesticides and from people who do not have direct contact with these substances to define the negative control group. The study observed a significant induction in the increase of micronuclei, thus showing an increase in the levels of genotoxicity in rural workers.

In the work with the fungicide Tebuconazole, in the micronucleus test on human T lymphocytes, the pesticide showed mutagenic activity. When the mean concentrations were compared to the control group, there was a significant difference, and it also indicated a correlation between the concentrations, showing a dose-dependency, and as the concentration of the fungicide increased, so did the mean number of micronuclei. The highest concentration of 100 per cent (10pl) showed the highest average number of micronuclei.

This corroborates the work of De OLIVEIRA, PC; MEIRELES, JRC (2011), who carried out a study with agricultural workers in the municipality of Ponto Novo - BA. The study did not classify the pesticides used in the region, but carried out micronucleus tests on the oral mucosa of the workers, people living around the irrigation areas of the plantations and unexposed people. The study also analysed other genetic alterations such as the presence of karyorrhexis, karyolysis, pyknosis, condensed chromatin and broken-eggs.

The results showed that the frequency of micronuclei was higher in the exposed group (EG) than in the control group (CG), and the frequency of condensed chromatin and caryorrhoea was also higher in the EG than in the CG, with a significant difference. When compared to the population living in the irrigated perimeter, there was no significant difference in relation to the frequency of micronuclei, but in relation to the other alterations analysed there was a significant difference. Thus, the study found that in relation to the control group for agricultural workers there was a significant difference in relation to the frequency of micronuclei and in relation to the other alterations analysed, while comparing agricultural workers with residents of the irrigated perimeter, there was a higher frequency of micronuclei but no significant difference, but in relation to the other alterations there was a significant difference, which may indicate a toxic potential of these pesticide mixtures.

In a study carried out in the village of Vila Bessa in the municipality of Conceição do Jacuípe - Bahia (RAMOS, MESP., 2009), an analysis was carried out on individuals occupationally exposed to pesticides and other types of agrotoxins. In the study in question, various tests were carried out, but in the micronucleus test carried out, the exposed group had a higher frequency of micronuclei than the control group, but there was no significant difference between them. However, in other tests there were significant differences.

Corroborating the work of De OLIVEIRA, sNK; FERREIRA, LG (2008), this study was carried out in the state of Goiás, in the municipalities of Portelândia and Palestina de Goiás, with occupationally exposed individuals. The micronucleus test was carried out on cells from the oral mucosa and analysed. A significant difference was found between the exposed group and the control group, with a higher frequency of micronuclei in the first group.

In the study by VENTURA, BC et. al. (2004), *micronucleus* tests were carried out using the *Oreochromis niloticus* test system, where they were exposed to concentrations of the herbicide Atrazine. The study found that the herbicide showed a significant difference in relation to the negative control, and the concentrations of Atrazine showed differences between them, with the higher the concentration the higher the frequency of micronuclei and other alterations.

Corroborating with the work of GOLDONI, A; da SILVA, LB (2012) where work was carried out with the fungicide Mancozeb, with the test system *Astyanax jacuhiensis*, and the micronucleus test was carried out by collecting blood from the tail vein, there was a higher frequency of micronuclei in the exposed group but there was no significant difference when compared to the control group.

In the study by SOUZA - FILHO, J (2011), the herbicide Roundup Transorb was tested on Guppy fish (*Poecilia reticulata*), and nuclear alterations and micronuclei were assessed, where it was observed that the exposed group showed a significant difference when compared to the control group, and alterations also increased as the concentration of the herbicide increased.

CHAPTER 5

CONCLUSION

As has been discussed throughout the text, the use of pesticides, especially in Brazil, is very intense and on a large scale, with the country being one of the biggest consumers of pesticides, given the large volume of grains and other foods that Brazil consumes and also exports. However, even with the supervision of the responsible bodies, there are still many irregularities in the use and application of these agrochemicals, as shown by the fact that illegal substances and agrochemicals have been incorporated into food, as well as the fact that there are residues of agrochemicals that are higher than permitted in various types of food.

The indiscriminate use of pesticides can cause damage to human health as well as to the environment. Poisoning from these compounds, which are often highly toxic, can also accumulate in the body and cause mutations in our cells, which can then lead to various types of problems and illnesses, such as neuronal diseases and cancer.

Several studies have been carried out on the mutagenicity, cytotoxicity and genotoxicity of various pesticides, using plant models such as Allium cepa (onion), and tests on animal cells that can assess chromosomal and nuclear aberrations and micronuclei (MN), which can be assessed in vivo or in vitro. As we can see, these tests have great sensitivity in showing us the high toxic capacity of pesticides and their components. Several studies have reported the genotoxicity, mutagenicity and cytotoxicity of pesticides in these cells, suggesting that greater care should be taken when using these components.

This study obtained results showing that the fungicide Tebuconazole had mutagenic and cytotoxic potential in the Allium cepa test, but the 100 per cent concentration was the one with the lowest average number of alterations, which could be caused by the cytotoxic potential of the fungicide, which impaired cell division, among other reasons, but the 75 per cent concentration was the one with the highest average number of alterations, indicating that if the doses were higher, we would possibly have a higher average number of alterations.

In the micronucleus test, Tebuconazole also showed mutagenic and cytotoxic potential, when comparing the exposed group to the negative control group, there was a significant difference, in this test also showing a dose-dependent correlation of the fungicide, since as the concentration increased, the micronucleus averages also increased.

It can be concluded from this study that the fungicide Tebuconazole has a mutagenic and cytotoxic character, and can be harmful to the health of people who have worked with it, or have some contact with the pesticide, and also harmful to the environment and all living beings present in that ecosystem that is being affected by the fungicide, In order for workers to avoid health problems, it is recommended that they wear PPE (personal protective equipment) and take greater care when applying the pesticide to avoid contaminating the soil, water, vegetation and other areas.

Bibliographical references

AMBROSIO, J.B; FERNANDES, T.C; MARIN-MORALES, M.A. Cytotoxic, genotoxic and mutagenic effects of the insecticide imidacloprid, the herbicide sulfentrazone and a mixture of these two pesticides on Allium cepa cells. Paulista State University. 2012.

AMBROSIO, J.B; CABRAL-DE-MELO, D.C; MANTOVANI, M.S; MARIN-MORALES, M.A. Genotoxic and mutagenic action of environmental concentrations of the insecticide imidacloprid, the herbicide sulfentrazone and the combination of these two pesticides on human cells in vitro and on Salmonella typhimurium. Paulista State University. 2012.

ANDRADE, L. F; DAVIDE, L. C; GEDRAITE, L.S. The effect of cyanide compouds, fluorides and inorganic oxides present in spent pot Linner on germination and root tip cells of Lactuta sativa. Ecotoxicology and Environmental Safey, v. 25, n. 73, p. 626-631. 2010.

AQUINO, I. **Genotoxic effect of artemisinin and artesunate on mammalian cells.** 2010. 81f. Master's thesis. Universidade Estadual **Paulista "Julio de Mesquita Filho", Botucatu, São Paulo. 2010.**

AU, W.W.; WARD JR., J.B.; RAMANUJAM, V.M.S.; HARPER, B.L.; MOSLEN, M.T.; LEGATOR, M.S. Genotoxic effects of a sub-acute low-level inhalation exposure to a mixture of carcinogenic chemicals. **Mutation Research**, v. 203, p. 103-115. 1988.

BARKY, F. A, ABDELSALAM, H. A, MAHMOUD, M. B, HAMDI, S. A. H. Influence of atrazine and Roundup pesticides on biochemical and molecular aspects of Biomphalaria alexandrina snails. **Pestic Biochem Physiol**, v. 104, p. 9-18. 2012.

Bayer CropScience Limited, 2005.

BELDEN, J.B.; LYDY, M.J. Impact of atrazine on organophosphate insecticide toxicity. **Environmental Toxicology and Chemistry**, v. 19, p. 2266-2274. 2000.

BELPOGGI, F. et al. Results of Long-Term Experimental Studies on the Carcinogenicity of Ethylene-bis-dithiocarbamate (Mancozeb) in **Rats. Ann. New York Acad. Sci.** New York, v. 982, p. 123- 136, Jan. 2002.

BENDER, R. P. et al. Polychlorinated Biphenyl Quinone Metabolites Poison Human Topoisomerase II Alpha: Altering Enzymefunction by Blocking The N-

Thermal Protein Gate. **Biochemistry**, New York, v. 45, n. 33, p. 10140-10152, Sep. 2006.

BERNARDES, P. M et *al.* **Prospection of fungicide toxicity by macroscopic, microscopic and molecular analyses in Allium cepa**. Dissertation submitted to the Postgraduate Programme in Plant Production. Federal University of Espírito Santo. 2014.

BERNARDI, D. **Mutation and repair**. 2011.

BIANCHI-SANTAMARIA, A.; GOBBI, M.; CEMBRAN, M.; ARNABOLDI, A. Human lymphocyte micronucleus genotoxicity test with mixtures of phytochemicals in environmental concentrations. **Mutation Research**, v.388, p. 27-32. 1997.

BICKHAM, J. W; SANDHU, S; HEBER, P. D. N; CHIKHI, L; ANTHWAL, R. Effects of chemical contaminants on genetic diversity in natural populations: implications for biomonitoring and ecotoxicolgy. **Mutation Research**, v. 463, p. 33 - 51.2000.

BIGGER, A.H.; CLARKE, J.J. L5178Y TK +/- Mouse lymphoma mutagenesis assay with a confirmatory assay. Rockville, MD: **Microbiological Associates**, Inc. Laboratory Study No. 1992.

BINDALI, B. B.; KALIWAL, B. B. Anti-Implantation Effect of a Carbamate Fungicide Mancozeb in Albino Mice. **Ind. Health**, v. 40, p. 191-197, Jan. 2002.

BISSON, M.; HONTELA, A. Cytotoxic and Endocrine-Disrupting Potnetial of Atrazine, Diazinon, Endosulfan and Mancozeb in Adrenocortical Steroidogenic Cells of Rainbow Trout Exposed in vitro. **Toxicol. Appl. Pharmacol.**, v. 180, p. 110-117, May 2002.

BOLLE, P. S, et al. Clastogenicity of atrazine assessed with the Allium cepa test. **Environmental Molecular Mutagenesis**, v. 43, p. 137-141. 2004.

BOLOGNESI, C; BONATTI, S; DEGAN, P; GALLERANI, E; PELUSO, M; RABBONI, R; ROGGIERI, P; ABBONDANDOLO, A. Genotoxic activity of glyphosate and its technical formulation Roundup. **J Agric. Food Chem**, v. 45, p. 1957-1962. 1997.

BOLOGNESI, C.; MORASSO, G. Genotoxicity of pesticides: potential risk for consumers. **Trends in Food Science and Technology**, v. 11, p. 182-187. 2000.

BOLOGNESI, C. Genotoxicity of pesticides: a review of human biomonitoring studies. **Mutation Research**, v. 543, p. 251-272, 2003.

BRAGUINI, W. L. **Effects of Deltamethrin and Glyphosate on Parameters of Mitochondrial Energy Metabolism, on Artificial and Natural Membranes in "in vivo" Experiments.** 2005, 191 f. Dissertation (Doctorate in Biochemical Sciences) - Federal University of Paraná, Curitiba, 2005.

CABALLO, C. et al. Analysis of Cytogenetic Damage Induced in CHO Cells by the Pyrethroid Insecticide Fenvalerate: Teratogen, Carcinogen, **Mutagen. Toxicology**, Houston, v 12, n. 6, p. 243-249, Sep. 1992.

CABRERA, GL; RODRIGUEZ, DMG. Genotoxicity of soil from farmland irrigated with wastewater using three plant biosays. **Mutation Research**. v. 426, p. 211-214.1999.
CARTER, S. B. Effects of cytochalasyn on mammalian cells. **Nature**, v. 213, p. 261-264. 1967.

CALIFORNIA EPA DEPARTMENT OF PESTICIDE REGULATION, USA. 2006.

CARVALHO, M.B et. al. Correlation between clinical evolution and the frequency of micronuclei in cells from patients with oral and oropharyngeal carcinomas. **Revista Assoc Med Bras**. v. 48, n. 4, p. 317-322. 2002.

CASTRO, V.L.S.S.; DESTEFANI, C.R.; DINIZ, C.; POLI, P. Evaluation of neurodevelopmental effects on rats exposed prenatally to sulfentrazone. **Neurotoxicology**, v. 28, p. 1249-1259, 2007.

CAVALCANTE, D. G. S. M; MARTINEZ, C. B. R; SOFIA, S. H. Genotoxic effects of Roundup on the fish Prochilodus lineatus. **Mutation Research**. 2008.

CAVALLO, D.; URSINI, C.L.; FRESEGNA, A.M.; CIERVO, A.; MAIELLO, R.; RONDINONE, B.; D'AGATA, V.; IAVICOLI, S. Direct-oxidative DNA damage and apoptosis induction in different human respiratory cells exposed to low concentrations of sodium chromate. **Journal of Applied Toxicology**, v. 30, p. 218-225. 2010.

CHAUHAN, L. K. S, SAXENA, P. N, GUPTA, S. K. Cytogenetic effects of cypermethrin and fenvalerate on the root meristem cell of *Allium cepa.* **Environ Exp Bot**, v. 42, p. 181-189. 1999.

CHAVES, T.V.S. **Avaliação do impacto do uso de agrotóxicos nos trabalhadores rurais dos municípios de Ribeiro Gonçalves, Baixa Grande do Ribeiro e Uruçuí - Piauí.** Postgraduate dissertation in Pharmacology, Federal University of

Ceará.2007.

CLEMENTS, C; RALPH, S; PETRAS, M. Genotoxicity of selected herbicides in Rana catesbiana tadpoles using the alkaline single-cell gel DNA electrophoresis (comet). **Environ. Mol. Mutagen.** V. 29, p. 277-288. 1997.

COGERH. Participatory Aquifer Management Plan for the Potiguar Basin, State of Ceará - Final Report. Fortaleza, 2009.

COSTA, C.; SILVARI, V.; MELCHINI, A.; CATANIA, S.; HEFFRON, J.J.; TROVATO, A.; DE PASQUALE, R. Genotoxicity of imidacloprid in relation to metabolic activation and composition of the commercial product. **Mutation Research**, v. 672, p. 40-44. 2009.

COSTA, R.M.A; MENK, C.F.M. Biomonitoring of environmental mutagenesis. **Biotecnologia: Ciência & Desenvolvimento**. v.2, n.12, p24-26, 2000.

COX, C. Glyphosate (Roundup). **Journal of Pesticide Reform**, v. 18, p. 3-17. 1998.

ÇAVAS, T; KONEN, S. Detection of cytogenetic and DNA damage in periphenal erythrocytes of goldfish (Carassius auratus) exposed to a glyphosate formulation using the micronucleus test and the comet assay. **Mutagenesis**, v. 22, p. 263-268. 2007.

DAS, P.P.; SHAIK, A.P.; JAMIL, K. Genotoxicity induced by pesticide mixtures: in-vitro studies on human peripheral blood lymphocytes. **Toxicology and Industrial Health**, v. 7, p. 449-458. 2007.

De OLIVEIRA, P.C.; MEIRELES, J.R.C. **Cytogenetic biomonitoring of agricultural workers from the irrigated perimeter in the municipality of Ponto Novo - BA**.Final course work.Universidade Estadual de Feira de Santana.2011.

De OLIVEIRA, S.N.K; FERREIRA, L.G. **Cytogenetic analysis of individuals occupationally exposed to pesticides**. Abstract presented at the State University of Goiás scientific initiation event. 2008.

De ROBERTIS, E. M. F.; HIB, J. and PONZIO, R. **DNA replication**, In: Biologia celular e molecular. Guanabara Koogan Publishing House. Rio de Janeiro, RJ, 2003.

DEMSIA, G.; VLASTOS, D.; GOUMENOU, M.; MATTHOPOULOS, D. P. Assessment of the genotoxicity of imidacloprid and metalaxyl in cultured human lymphocytes and rat bonemarrow. **Mutation Research**, v. 634, p. 3239. 2007.

DELLA ROSA, HENRIQUE V; SIQUEIRA, MARIA E. P; COLACIOPPO,

SERGIO. Environmental and Biological Monitoring. **In: OGA, Seizi. Fundamentals of Toxicology**. 2.Ed. São Paulo: Atheneu, p.145-161. 2003

DIETZ, J. Micronucleus research in oesophageal mucosa and its relationship with risk factors for oesophageal cancer. **Rev Ass Med Brasil**. v.46, n.3, p. 207211. 2000.

DIMITROV, B. D; GADEVA, P. G; BENOVA, D. K; BINEVA, M. V. Comparative genotoxicity of the herbicide roundup, stomp and reglone in plant and mammalian test systems. **Mutagenesis**, v. 21, p. 375-382. 2006.

DISNER, G.R. et. al. Evaluation of the mutagenic activity of Roundup® in Astyanax altiparanae (Chordata, Actinopterygii). **Evidência,** Joaçaba. v. 11, n. 1, p. 33-42, January/June 2011.

DOLARA, P.; SALVADORI, M.; CAPOBIANCO, T.; TORRICELLI, F. Sister chromatid exchanges in human lymphocytes induced by dimethoate, omethoate, deltamathrin, benomyl and their mixture. **Mutation Research**, v. 283, p. 113-118. 1992.

ELIK, A. C. et al. Evaluation of Cytogenetic Effects of Lambda-Cyhalothrin on Wistar Rat Bone Marrow by Gavage Administration. Ecotoxicol. Environ. Saf. **Charlottesville**, v. 61, n.7, p. 128-133, Oct. 2005.

ENAN, M. R. Assessment of genotoxic activity of Para-nitrophenol in higher plant using arbitrarily primed-polymerase chain reaction (AP-PCR). **American Journal of biotechnology and biochemistry**, v. 3, n. 2, p. 103 - 109. 2007.

EPA; US EPA, Pesticide Registration Status. 2006.

EVANS, H.J. Neoplasia and cytogenetic abnormalities. In: DELLARCO, V.L.; VOYTEK, P.E.; HOLLAENDER, A. (Ed). **Aneuploidy etiology and mechanisms**. New York: Plenum Press, p. 165-178. 1985.

FAO - **Food and Agriculture Organisation of the United Nations. New Code of Conduct on Pesticides adopted**. Rome, FAONEWSROOM, 4 Nov. 2002.

FENECH, M; MORLEY A. A. Solutions to the kinetic problem in the micronucleus assay. **Cytobios,** n. 43, p. 233- 246. 1985.

FENECH, M. The in vitro micronucleus technique. **Mutation Research**. n.455, p81- 95, 2000.

FENG, S.; KONG, Z.; WANG, X.; ZHAO, L.; PENG, P. Acute toxicity and genotoxicity of two novel pesticides on amphibian, Rana N. Hallowell. **Chemosphere,** v. 56, p. 457-463. 2004.

FENG, S; KONG, Z; WANG, X; PENG, P; ZENG E.Y. Assessing the genotoxicity of imidacloprid and RH-5849 in human peripheral blood lymphocytes in vitro with comet assay and cytogenetic tests. **Ecotoxicol. Environ. Saf**. V. 61, p. 239-246. 2005.

FENSKE, R.A. Pesticide exposure assessment of workers and their families. **State of the Art Reviews Occupational Medicine**, v. 12, p.221-237. 1997.

FERNANDES, T.C.C. et al. **Investigation of the toxic, mutagenic and genotoxic effects of the herbicide Trifluralin, using** Allium cepa **and *Oreochromis niloticus* as test systems.** Dissertation presented to the Biosciences Institute of the Rio Claro Campus, Paulista State University, for the master's degree course in Biological Sciences. 2005.

FERREIRA, DAC; MOTTA, LC; KREUTZ, C; TONI, VL; LORO; LJG, BARCELLOS. Assessment of oxidative stress in *Rhamdia quelen* exposed to agrichemicals. **Chemosphere**, v. 79, p. 914-921. 2010.

FIDELES, N. Impacts of the Green Revolution. **RadioagenciaNP**, São Paulo, September 2006.

FISHEL, F. M. Pesticide Toxicity Profile: Synthetic Pyrethroid Pesticides. **Ecotoxicol. Environ. Saf**., Charlottesville, v. 60, n.5, p. 118-123, July. 2005.

FISKEJÓ, G. The *Allium* test as a standard in the environmental monitoring. **Hereditas**. v. 102, p. 99-112. 1985

FISKEJÓ, G. *Allium* Test II: Assesment of chemicaPs genotoxic potential by recording aberrations in chromosomes and cell divisions in root tips of *Allium cepa* L. **Environmental Toxicology and Water Quality**. v. 9, p. 235-24.1994.

FRANCO NETTO, G. On the need to assess cancer risk in populations environmentally and occupationally exposed to virus and chemical agents in developing countries. **Cadernos de Saúde Pública**. V. 14, p. 87 - 98, suppl. 3. 1998.

FRENZILLI, G; SCARCELLI, V; BARGA, I; NIGRO, M; FORLIN, L;BOLOGNESI, C; STURVE, J. DNA damage in eelpout (Zoarces viviparus) from Goteborg harbour, **Mutation Research**, v. 552, p. 187 - 195, 2004.

FUNDACENTRO. Prevention of accidents at work with pesticides: segurança e saúde no trabalho, n. 3. São Paulo: Fundação Jorge Duprat Figueiredo de Segurança e Medicina do Trabalho, Ministério do Trabalho, 1998.

GADANO, A; GURNI, A.; LÓPEZ, P.; FERRARO, G.; CARBALLO, M. Genotoxicity of medicinal plant infusions. **Revista Brasileira de Farmacognosia**.v. 17, n.3. 2002.

GARAJ-VRHOVAC, V.; ZELJEZIC, D. Assessment of genome damage in a population of Croatian workers employed in pesticide production by chromosomal aberration analysis, micronucleus assay and comet assay. **Journal of Applied Toxicology**, v. 22, p. 249- 255. 2002.

GOLDONI, A.; SILVA, L. B. Mutagenic potential of the fungicide Mancozeb in Astyanax jacuhiensis (Teleostei: Characidae). Biosci J. v. 28, n.2, p. 297-301. 2012.

GRANT, W. F. Chromosome aberrations assay in A report of the U.S. Environmental Protection Agency Gene- Tox Programme. **Mutation Res**, v. 99, p. 273-291. 1982.

GRIFFITHS, A. J. F.; GELBART, W. M.; MILLER, J. H.; LEWONTIN, R. C. Gene mutations, p. 177-208. In: **Modern Genetics**. Editora Guanabara Koogan, Rio de Janeiro, RJ. 2001.

GRIFFITHS, A. J. F; WESSLER; LEWOTIN; MILLER. **Introduction to Genetics**, ed. 8, p. 743, Editora Guanabara. 2006.

GRISOLIA, C.K. A comparison between mouse and fish micronucleus test using cyclophosphamide, mitomycin C and various pesticides. **Mutantion Research**, v. 518, p. 145-150. 2002.

GRISOLIA, C.K. Agrotoxins: **Mutations, cancer and reproduction**.Editora UNB.p. 63 - 75. 2005.

HARASHIMA, H; SCHNITTGER, A. The integration of cell division, growth and differentiation. **Current Opinion in Plant Biology**. 2010.

HODGSON, E. Induction and inhibition of pesticide-metabolising enzymes: roles in synergism of pesticides and pesticide action. **Toxicology and Industrial Health**, v. 15, p. 6-11. 1999.

HOGSTEDT, B. Micronuclei in lymphocytes with preserved cytoplasm. A method for assessment of cytogenetic damage in man. **Mutation Res**, 130: p. 63-72. 1984.

HOPPIN, J.A.; ADGATE, J.L.; EBERHART, M.; NISHIOKA, M.; RYAN, P.B. Environmental exposure assessment of pesticides in farmworker homes. **Environmental Health Perspectives**, v.114, p.929-935. 2006.

ISKANDAR, O. An improved Method for the Detection of Micronucleus in Human Lymphocytes. **Stain Technol.**, 54, p. 221-223. 1979.

JARRARD, H. E. et al. Impacts of Carbamate Pesticides on Olfactory

Neurophysiology Cholinesterase Activity in Coho Salmon (Oncorhynchus Kisutch). **Aquat. Toxicol**. Boston, v. 69, p. 133 - 148, Jan. 2004.

JEMEC, A; TISLER, T; DROBNE, D; SEPCIC, K; FOURNIER, D; TREBSE, P. Comparative toxicity of imidacloprid, of its commercial liquid formulation and diazinon to a non-target arthropod, the microcrustacean Daphia magna. **Chemosphere**, v. 68, p. 1408-1418. 2007.

JHA, A. N. Genotoxical studies in aquatic organisms: an overview. **Mutation Research**, v. 552, p. 1 - 17. 2004.

JIRAUNGKOORSKUL, W.; UPATHAM, E. S.; KRUATRACHUE, M.; SAHAPHONG, S.; VICHASI-GRAMS, S. & POKETHITIYOOK, P. Histopathological effects of Roundup, a glyphosate herbicide, on nile tilapia (Oreochromis niloticus). **Science Asia.** v. 28, p. 121-127. 2002.

KARABAY, N. U; OGUZ, M. G. Cytogenetic and genotoxic effects of the insecticides, imadocloprid and methamidophos. **Genetic. Mol. Res.**, v. 4, p. 635-662. 2005.

KHAN, P. K ; SINHA, S. P. Ameliorating Effect of Vitamin C on Murine Sperm Toxicity Induced by Three Pesticides (Endosulfan, Phophoamison and Mancozeb). **Mutagenesis**, London, v. 11, p. 33 - 36, July 1996.

KOIFMAN, R. J; MEYER, A. Human reproductive system disturbances and pesticide exposure in Brazil. **Cadernos de Saúde Pública**. V. 18, p. 435 - 445 n. 2, mar-abr, 2002.

KREUTZWEISER, D. P; GOOD, K; CHARTRAND, D; SCARR, T; THOMPSON, D. Non-target effects on aquatic decomposer organisms of imidacloprid as a systemic insecticide to control emeral ash borer in riparian trees. **Ecotoxicol. Environ. Saf**. V. 68, p. 315-325. 2007.

KRUGER, R. A et al. **Analysis of the toxicity and genotoxicity of pesticides used in agriculture using bioassays with** Allium cepa. Dissertation presented to the Environmental Quality Programme at the Feevale University Centre. 2009.

LEME, D. M.; MARIN-MORALES, M. A. Allium cepa Test in environmental monitoring: a review on its application. **Mutation Research**, Amsterdam, v. 682, p. 71-81, 2009.

LEWIN, B. Genes VII. Porto Alegre: **Artmed**, p. 990. 2001.

LIOI, M. B; SCARFI, M. R; SANTORO, A; BARBIERI, R; ZENI, O; SALVEMINI, F; DI BERNARDINO, D; URSNI, M. V. Cytogenetic damage and induction of pro-oxidant state in human lymphocytes exposed in vitro to gliphosate, vinclozolin, atrazine and DPX-E9636. **Environmental and Molecular Mutagenesis**, v. 32, p. 39-46. 1998a.

LIOI, M. B; SCARFI, M. R; SANTORO, A; BARBIERI, R; ZENI, O; SALVEMINI, F; DI BERNARDINO, D; URSNI, M. V. Genotoxicity and oxidative stress induced by pesticide exposure in bovine lymphocyte cultures in vitro. **Mutation Research**, v. 403, p. 13-20. 1998b.

LIU, P.J.W; LI, X.M; QI, Q.X; ZHOU, L; ZHENG, T.H; SUN, Y.S. YANG. DNA changes in barley (Hordeum vulgare) seedlings induced by cadmium pollution using RAPD analysis. Chemosphere, v. 61, p. 158 - 167. 2005.

LIU, W; YANG, Y. S; LI, P. J; ZHOU, Q. X; XIE, L. J; HAN, Y. N. Risk assessment of cadmium-contaminated soil on plant DNA damage using RAPD and physiological indices. **Journal of Hazardous Materials**, v. 161, p. 878883. 2009.

LIVINGSTONE, D.R. Contaminant-stimulated reactive oxygen species production and oxidative damage in aquatic organisms. **Bulletin of Marine Pollutants**. v. 42, p. 656 - 666. 2001.

LYONS, B.P.; HARVEY, J.S.; PARRY, J.M. An initial assessment of the genotoxic impact of the Sea Empress oil spill by the measurement of DNA adduct levels in the intertidal teleost Lipophrys pholis. **Mutation Research**, v.390, p.263-268, 1997.

MA, T. H. et al. The Improved Allium/Vicia Root Tip Micronucleus Assay for Clastogenicity of Environmental Pollutants. **Mutation Research**, Orlando, v. 334, n.5, p. 185 -195, Oct.1995.

MARINHO, AP. **Contexts and risk contours of agricultural modernisation in the municipalities of Baixo Jaguaribe-Ce: the mirror of (dis)involvement and its effects on health, work and the environment.** PhD Thesis, School of Public Health/USP, 2010.

MATSUMOTO, S.T. et al. 2006. Genotoxicity and mutagenicity of water contaminated with tannery effluents, as evaluated by the micronucleus test and comet assay using the fish Oreochromis niloticus and chromosome aberrations in onion root-tips.**Genetics and Molecular Biology**, 29: 148-158. 2006.

MIRANDA, A. C.; MOREIRA, J. C.; CARVALHO, R.; PERES, F. Neoliberalism, the use of pesticides and the crisis of food sovereignty in Brazil. **Ciênc. Saúde Coletiva**, Rio de Janeiro, v.12, n. 1, p. 15-24. 2007.

MIRANDA, C.T. **Ecotoxicological assessment of lakes in the Goiânia region.** Monograph presented to the Biology Department of the Pontifical Catholic University of Goiás. Goiânia-Goiás. 2009.

MOHANTY, G. J; MOHANTY, S. K; GARNAYAK, DUTTA. Report on wildlife poisoning by Furadan. **Widl Direct Kenya**, v. 33, p. 771 - 780. 2009.

MONSERRAT, J.M; MARTINEZ, P.E; GERACITANO, L.A; AMADO. L.L; MARTINS, C.M; PINHO, G.L; CHAVES, I.S; FERREIRA-CRAVO, M; VENTURA-LIMA, J; BIANCHINI, A. Pollution biomarkers in estuarine animals: critical review and new perspectives. Comparative Biochemistry and Physiology part C: Toxicology and Pharmacology, v. 146, n. 1-2, p. 221-34. 2007.

MOREIRA, R. J. **Environmentalist criticisms of the Green Revolution**. In: WORLD CONGRESS OF RURAL SOCIOLOGY-IRSA, 10th; BRAZILIAN CONGRESS OF RURAL ECONOMIC AND SOCIOLOGY-SOBER, 37th, 2000, Rio de Janeiro.

MOREIRA, L.M.A., ARAUJO, L.M.P., CORDEIRO, A.P.B., GUSMAO, F.A.F. Test of human lymphocytes in the recognition of the clastogenic and cytotoxic effect of 5- fluoracil. R. Ci. **Med. biol**. v.3, n.1, p.5-12, 2004.

MONTEIRO, D.A. ALVES DE ALMEIDA, J. RANTIN, F.T. & KALININ, A.L. Oxidative stress biomarkers in the freshwater characid fish, Brycon cephalus, exposed to organophosphorus insecticide Folisuper 600 (methyl parathion). **Comparative Biochemistry and Physiology**, Part C v. 143, p. 141-149. 2006.

MURLI, H. Mutagenicity test on F6285 technical, PL96-011, in a dominant lethal assay in rats. Vienna, VA: Corning Hazleton Inc. (CHV), CHV Study No. 17564-0-472, FMC Study No. A96-4429. 1996.

M. VANHANEN, T. TUOMI, U. TIIKKAINEN et al. Sensitisation to enzymes in the animal feed industry. **Occupational and Environmental Medicine**, vol. 58, no. 2, pp. 119-123, 2001

NESKOVIC, N.K.; POLEKSIC, V.; ELEZOVIC, I; KARAN, V.; BUDIMIR, M. Biochemical and histopathological effects of glyphosate on carp (Cyprinus carpio). **Bulletin of Environmental Contamination and Toxicology**. v.56, p. 295-302. 1996.

NEVES, J.T. **Evaluation of genetic damage in lymphocytes of workers exposed to pesticides through the micronucleus test**.Master's thesis in Environmental Contamination and Toxicology; Faculty of Sciences, University of Porto. 2010.

NONDILLO, M. et al. Effect of Neonicotinoid Insecticides on the South American Fruit Fly Anastrepha Fraterculus in Grapevine Culture. **BioAssay**, Bento

Gonçalves, v. 2, n. 9, p. 5-14, Feb. 2007.

NUNES, W. B. **Evaluation of the mutagenic and/or recombinogenic potential of cottonseed in somatic and germ cells of D. melanogaster**. 2000. 112f. Master's dissertation. Federal University of Goiás. Goiânia. 2000.

OGA, S. **Fundamentals of toxicology**. 5 ed. São Paulo. Atheneu. 1996.

OLGUN, S.; GOGAL Jr, R.M.; ADESHINA, F.; CHOUDHURY, H.; MISRA, H.P. Pesticide mixtures potentiate the toxicity in murine thymocytes. **Toxicology**, v. 196, p. 181-195. 2004.

OLIVEIRA, L. M; VOLTOLINI, J. C; BARBÉRIO, A. Mutagenic potential of pollutants in the water of the Paraíba do Sul River in Tremembé, SP, Brazil, using the *Allium cepa* test. **Ambi-Agua**, Taubaté, v. 6, n. 1, p. 90 - 103. 2011.

PENA, R. V et al. Toxicological genetic effect of acute exposure of Poecilia reticulata (Poecilidae) to the fungicide tebuconazole. Paper presented at the Brazilian Congress of Genetics. 2011.

PERES, F.; MOREIRA, J. C.; DUBOIS, G. S. Pesticides, health and the environment: an introduction to the theme. In: PERES, F. and MOREIRA. J. C. (org). Is it poison or is it medicine? Pesticides, health and the environment. Rio de Janeiro: **Ed. FIOCRUZ**. 2003.

PERES, F.; ROZEMBERG, B.; LUCCA, S.R. Perception of risks in rural work in an agricultural region of the State of Rio de Janeiro, Brazil: pesticides, health and environment. **Caderno da Saúde Pública**. Rio de Janeiro, v. 21, p. 18361844 2005.

PUTNAM, D.L; YOUNG, R.R. Micronucleus cytogenetic assay in mice. Bethesda, MD: **Microbiological Associates**, Inc. Laboratory Study No. TA136.122019. 1992.

QUEIROZ EK, WAISSMANN W. Occupational exposure and effects on the male reproductive system. **Caderno de Saúde Pública**, v. 22, p. 485-93. 2006.

RABELLO-GAY, M. N.; RODRIGUES, M. A. R.; MONTELEONE-NETO, R. (Ed.). Mutagenesis, Teratogenesis and Carcinogenesis: evaluation methods and criteria. Ribeirão Preto: **Brazilian Journal of Genetics**. 1991.

RAMEL, C. et al. ICPEMC - Inhibitors of mutagens and their relevance to carcinogenesis. **Mutation Research**, v.168, p.47-65, 1986.

RAMOS, M.E.S.P. **Genetic biomonitoring of individuals occupationally exposed to pesticides in the village of Vila Bessa, municipality of Conceição de Jacuípe, Bahia**.

PhD thesis presented to the Postgraduate Coordinator in Pharmacology, Faculty of Medicine, Federal University of Ceará. 2009.

RANK, J.; NIELSEN, M.H. Evaluation of the Allium anaphase-telophase test in relation to genotoxicity screening of industrial wastewater. **Mutation Research**. v. 312, n. 1, p. 17 - 24. 1993.

REGISTRATION: Ministry of Agriculture, Livestock and Supply (MAPA) No. 02606.

RIBEIRO, L. R. Micronucleus test in rodent bone marrow in vivo. **In: Environmental Mutagenesis**. Canoas: ULBRA, 2003.

RIBEIRO, L.R.; MARQUES, E.K. A importância da mutagenese ambiental na carcinogenese humana In: **Mutagênese Ambiental**. ULBRA Publishing House. Canoas-RS. 2003.

RIBEIRO, L.R.; SALBADORI, D.M.F.; MARQUES, E.K. **Environmental Mutagenesis**. ULBRA Publishing House. Canoas-RS. 2003.

RIGOTTO, RM et al. **Epidemiological study of the population of the lower Jaguaribe region exposed to environmental contamination in an area where pesticides are used**. Fortaleza. 2010.

RIGOTTO, RM. **Pesticides**. TRAMAS Centre - Work, Environment and Health for Sustainability Federal University of Ceará. 2012.

SABALIUNAS, D.; LAZUTKA, J.; SABALIUNIENE, I.; SODERGREN, A. Use of semipermeable membrane devices for studying effects of organic pollutants: comparison of pesticide uptake by semi-permeable membrane devices and mussels. **Environmental Toxicology and Chemistry**, v. 17, p. 1815-1824. 1998.

SALVADORI, D.M.F., RIBEIRO, L.R., FENECH, M. Micronucleus test in human cells in vitro **In: Environmental Mutagenesis**. Canoas: ULBRA, chap. 8 p. 201-223. 2003.

SAXENA, P. N; GUPTA, S. K; MURTHY, R.C. Carbofuran induced cytogenetic effects in root meristem cells of Allium cepa and Allium sativum: A spectrocospic approach for chromosome damage. **Pesticide Biochemistry and Physiology**, v. 96, p. 93-100. 2010.

SEVGILER, Y; ORUÇ, EO; UNER, N. Evaluation of etoxazole toxicity in the liver of *Oreochromis niloticus*. **Pestic. Biochem. Physiol**, v. 78, p. 1-8. 2004.

SIGRH: **Information System for Water Resources Management in the State of São Paulo**. Status Report on Water Resources in the Alto Paranapanema River Basin -

Report 1. Use of Pesticides in Agriculture. 2005.

SILVA, J. M.; NONATO-SILVA, E.; FARIA, H. P.; PINHEIRO, T. M. M. **Agrotóxico e Trabalho: uma combinação perigoso para a saúde do trabalhador rural.** Ciência & Saúde Coletiva. ABRASCO, v. 10. n. 4, p.891- 903. 2005.

SINDAG. **The Brazilian Plant Protection Products Market. Presented at the Workshop Evaluation of the Exposure of Mixers, Fillers and Applicators to Pesticides.** 2008.

National Toxic-Pharmacological Information System. **Annual statistics of poisoning cases**: Brazil. 2004.

Toxicity and genotoxicity of the herbicide Roundup Transorb evaluated in Guppy (Poecilia reticulata, **Cyprinodontiformes, Poeciliidae) in acute treatment**. 2011. Master's dissertation in Biological Sciences, Federal University of Goiás.

TAKAHASHI, C.S. In vitro cytogenetic tests and aneuploidy In: **Environmental Mutagenesis**. Canoas: ULBRA, chap. 6 p. 151-172. 2003.

TEIXEIRA, R.O.; CAMPAROTO, M.L.; MANTOVANI, M.S.; VICENTINI, V.E.P. Assessment of two medicinal plants, Psidium guajava L. and Achillea millefolium L. in in vivo assays. **Genet Mol Biol**. v. 26, p. 551-555. 2003.

THEODORAKIS, C.W.; BICKHAM, J.W.; ELBL, T.; SHUGART, L.R.; CHESSER, R.K. Genetics of radionuclide - contaminated mosquito fish populations and homology between Gambusia affins and G. Holbrooki. **Environmental Toxicology and Chemistry**, v.17, p.1992- 1998, 1998.

TISCH, M. et al. Genetoxic Effects of Pentachlorophenol, Lindane, Tranfluthrin, Cyfluthrin, And Natural Pyrethrum on Human Mucosal Cells Of The Inferior And Middle Nasal Conchae. **Am. J. Rhinol**. New York, v. 19, n. 29, p. 141-151, Dec. 2005.

TONI, CD; FERREIRA LC; KREUTZ, VL; LORO; LJG, BARCELLOS. Assessment of oxidative stress and metabolic changes in common carp *(Cyprinus carpia)* acutely exposed to different concentrations of the fungicide tebuconazole. **Chemosphere**, v. 83, p. 579-584. 2011.

UMBUZEIRO, G.A. et al. **Study of the genotoxicity of samples from the Ribeirão dos Cristais basin.CETESB.**São Paulo. 2003.

UMBUZEIRO, GA; ROUBICEK, DA. **Environmental genotoxicity**. Ecotox Aquática, chap. 14, p. 327 - 328. 2006.

VALERIO, M. E; GARCIA, J. F; PEINADO, F. M. Determination of phytotoxicity of soluble elements in soils, based on a bioassay with lettuce *(Lactuca sativa)*. **Science Total Enviroment**, v. 378, p. 63-66. 2007.

VENTURA, B.C et al. **Evaluation of the cytotoxic, genotoxic and mutagenic effects of the herbicide Atrazine, using** Allium cepa **and** Oreochromis niloticus **as test systems.**

Institute of Biosciences of the São Paulo State University "Julio de Mesquita Filho" for the master's programme in Biological Sciences. 2004.

VENTURA, B.C et al..**Evaluation of the mutagenic and genotoxic effects of the herbicide Atrazine, by means of the micronucleus and nuclear abnormality tests, using Oreochromis niloticus as a test system.** Dissertation presented to the Institute of Biosciences of the Universidade Estadual Paulista "Julio de Mesquita Filho" for the master's degree course in Biological Sciences. 2004.

VICENTINI, V.E.P., CAMPAROTO, M.L.; TEIXEIRA, R.O. MANTOVANI, M.S.. Averrhoa carambola L., Syzygium cumini (L.) Skeels and Cissus sicyoides L.: medicinal herbal tea.**Toxicology in vitro**. v. 18, n. 5, p. 617 - 622. 2001.

WHITE, P.A. and RASSMUSSEN, J.B. The genotoxic hazards of domestic wastes in surface waters. **Mutation Research**, 410: p. 223-236. 1998.

WLDEMAN, A. G; NAZAR, R.N. Significance of plant metabolism in the mutagenicity and toxicity of pesticides. Can. J. Genet. Cytol. V. 24, p. 437-449. 1982.

WINSTON, G. W.; DI GUILIO, R. T. Prooxidant and antioxidant mechanisms in aquatic organisms. **Aquat. Toxicol.**, v.19, p. 137-161, 1991.

YILDIZ, M; CIGERCI, I.H; KONUK, M; FIDAN, A. F; TERZI, H. Determination of genotoxic effects of copper sulfate and cobalt chloride in Allium cepa root cells by chromosome aberration and comet assays. **Chemosphere**, v. 75, p. 934-938. 2009.

Zhiyi R, Haowen Y. A method for genotoxicity detection using random amplified polymorphism DNA with Danio rerio. **Ecotoxicol Environ Saf**, v. 58, p. 96103. 2004.

Printed by Books on Demand GmbH, Norderstedt / Germany